REMUNERATIVE RAILWAYS

FOR NEW COUNTRIES;

WITH SOME ACCOUNT OF

THE FIRST RAILWAY IN CHINA.

BY

RICHARD C. RAPIER,

MEMBER AND TELFORD MEDALLIST INST. C.E.

NUMEROUS ILLUSTRATIONS AND ESTIMATES.

LONDON: E. & F. N. SPON, 46, CHARING CROSS.

NEW YORK: 446, BROOME STREET.

BERLIN: Asher & Co. ST. PETERSBURG: Watkins & Co.

CALCUTTA: Thacker, Spink, & Co. BOMBAY: Thacker, Vining, & Co.

MELBOURNE: G. Robertson.

1878.

CONTENTS.

—•◦✕◦•—

ILLUSTRATIONS.

(MOST OF THE WOODCUTS ARE FROM PHOTOGRAPHS.)

PHOTOGRAPHS.

PREFACE.

THE present work has arisen from the constant inquiries, received by the Author, from persons anxious to make railways in new countries, or in new districts of old countries. There is a general outcry, on the one hand, for more railways; and on the other hand, that many of the lines already made do not pay their way satisfactorily. The following pages have been written, in the hope of contributing somewhat towards a remedy; at any rate as regards future undertakings.

The practice of relying on guarantees, for the interest of capital, has to answer for much of the present stagnation and difficulty in introducing new works. In proposed railways abroad, it has only been too common, for a State to be asked to guarantee a dividend, upon as large a capital as the authorities could be induced to sanction. Large capital powers, fortified by a Government guarantee, are pretty sure to be exercised; and thus, in many cases, the expenditure has been quite out of proportion to the needs of the locality. As a natural consequence, the earnings of such railways have not been equal to the guarantee; and States, like individuals, soon become tired of guarantees when they are called on to pay them.

It seems probable that the next stage of railway development, throughout the world, will have to depend on the intrinsic merits of the undertakings, and their prospects of being able to earn their own living, rather than on any artificial support. It is therefore particularly opportune to inquire as to the practicability of introducing cheaper railways.

There is always considerable unwillingness to deviate from established types, and the desire to have as big and as good railways as anyone else, has led many into extravagance and mischief, both in this and in other countries.

During the last few years, much has been done towards facilitating the introduction of railways in the Colonies, and other distant countries, at greatly reduced cost as compared with the earlier railways. Not only has experience brought about the natural result of improved production, but a new departure altogether has been taken in constructing railways of a lighter calibre, than was formerly deemed practicable.

The Author has appended a table, giving various useful particulars, as to lines already constructed, both of the ordinary and the lighter types. These particulars are not as yet very complete, but the Author hopes, in subsequent editions, to be able to add to this store of useful information. Details as to actual earnings, cost of maintenance, &c., &c., are scarcely yet forthcoming from such lines as have already been made on the lighter system, but our information as to actual results will now rapidly increase, and the assistance of friends in supplying corrected particulars from time to time is cordially invited, and will be gratefully acknowledged in future editions.

It is hoped that the publication of such tabulated particulars from time to time will be of great service in showing :—

        1. What has already been done.

        2. At what cost ; and

        3. With what result.

Happily there is in the Engineering profession a constant recognition both of the duty and the policy of saving the client's money ; and, with this good spirit presiding, inquiry and ascertaining of facts must bear fruit to the real advantage of all.

In some of the chapters, incidental reference is made to the first railway in China. As that subject has become one of very general interest, the Author has given, in the form of an Appendix, a connected account of the various efforts which have been made, to add that vast empire to our field of engineering enterprise.

RICHARD C. RAPIER.

5, WESTMINSTER CHAMBERS, LONDON,
    December 21st, 1877.

REMUNERATIVE RAILWAYS

FOR

NEW COUNTRIES.

CHAPTER I.

THE total length of railway in Great Britain and Ireland, open for traffic at the end of 1876, was as follows :—

Single lines	7,703 miles.
Double „	8,889 „
Treble „	96 „
Quadruple	184 „
Total	16,872

Nearly all these lines are on the 4 feet 8¼ inch gauge, and are laid for the most part with rails of about 75 lbs. per yard, some of the main lines being laid with steel rails of 80 and 84 lbs. per yard. Their cost has been at the average rate of £39,000 per mile, including sidings and rolling stock (items which year by year add to the cost of the railway, by reason of demands for increased accommodation), and the average dividend on the cost has been at the rate of 4½ per cent. This dividend has been earned by an average receipt of £68 per mile per week, and 67 pence per train mile; the average working expenses being 55 per cent. of the gross receipts.

	Per Mile.
The English railways have been made at an average cost of ..	£39,000
Indian railways—Guaranteed lines	16,000
Indian State railways	9,000
Many colonial railways under	10,000
Some colonial railways as low as	7,000

Several steps further might be taken in the direction of cheaper railways, and very practicable lines may be made for £4000 per mile, and in some cases have been made for even less.

For the speeds now required on our main lines in Great Britain, the ordinary gauge of 4 feet 8¼ inches, with 80-lb. rail, may be considered as the least which can be expected to discharge the required duty with safety and certainty. There are, however, very many parts of England still in need of railway communication, which are not at all likely to demand any speed higher than 20 to 25 miles an hour.

B

For such localities much may be done by making railways about half the ordinary size; the word "size" is used because mere alteration of gauge will not make much difference—reduction to be of real use must be made throughout, especially in the weight of moving loads and in speed.

In this country the value of land is always a serious element in railway construction; and although there is no practical diminution in the quantity of land necessary for a narrow-gauge railway, as compared with an ordinary one, the sharper curves which are practicable, enable considerable saving to be effected, by laying out the line so as to avoid severance as much as possible, and by selecting the least expensive route, both as regards value of land and cost of works.

A light narrow-gauge line can be made at a saving of £3000 to £4000 per mile, as compared with the cost of the cheapest ordinary line, and when it is borne in mind that every thousand pounds per mile of cost requires £1 per mile per week of net revenue to pay 5 per cent. dividend, the importance of every saving is very striking; and when it is further taken into account that a light railway, with small trucks and engines, can be maintained and worked more cheaply than a full-sized line, a comparison somewhat as follows is soon arrived at.

Many local and branch railways earn no more than £10 to £12 per mile per week gross revenue, and as the traffic is usually carried on with engines weighing about 30 tons, heavy wagons, and carriages out of all proportion to the number of passengers conveyed, the outgoings are £9 or £10 per mile per week, leaving only £2 to £3 per mile per week as a net earning for a capital outlay of perhaps £10,000 to £15,000 per mile. Need it be wondered at that such lines are in a chronic state of difficulty, until they are at length swallowed up by some neighbouring large Company, at an enormous sacrifice to the shareholders of both undertakings !

If, on the other hand, the branch railway, earning £12 per mile per week, had been made on a smaller scale, capable of being worked with 10-ton engines, small carriages, and manageable wagons, its working expenses would probably be only £5 or £6 per mile per week, leaving a net revenue of £6 per mile per week. Now, such railways can be made even in England at a total cost of £5000 per mile, and the above net revenue would be 6 per cent. on this amount. There is no question as to the smaller railway being able to carry a traffic of four or five times the above amount, even on a single line of rails, and thus, from a revenue point of view, everything is to be gained by making the smaller railway at the lesser outlay.*

The great bugbear in treating of this subject is "break of gauge." Let it then be examined. First, with reference to passenger traffic, it may be remarked that, in nearly all cases, passengers have to change carriages at the junction of the branch with the main lines,

---

* The Isle of Man Railway, made on a gauge of 3 feet, with 40-lb. rails, and at a total cost of £6500 per mile, is worked at 41 per cent. of its gross receipts. The Ballymena and Red Bay Railway, on the same gauge, is worked at 33 per cent. of the gross receipts. The cost of working the Westleigh Railway, 3 feet gauge, 30-lb. rails, trucks 30 cwt. laden, is less than 50 per cent. of the receipts.

even though they be of the same gauge; and here it may be added that, in many cases "through" carriages are but very sparsely occupied, and the practice of running them is very much the reverse of economical.

Next as to goods. In such circumstances as the junctions of the narrow with the broad gauge railways at Gloucester, Bristol, and Exeter, break of gauge has been found in the highest degree inconvenient, owing to the great extent of the goods traffic; but still more to the fluctuations in the daily quantities requiring to be transferred, and the consequent uncertainty as to the amount of space, appliances, and labour required for accomplishing the transfer expeditiously. It is certain that a long system of railways with a break of gauge must always be at a disadvantage when it is in competition with other systems of continuous gauge; but the case of short branch lines, of a purely branch or feeder nature, is essentially different. On such lines, heavy fluctuations in the quantity of traffic to be transferred are not likely, and on a branch of say 10 miles in length the additional staff required for transhipment purposes would not exceed a cost of £8 or £10 per week, or about £1 per mile per week; whereas the saving to be gained by the lighter railway is at least £3 per mile per week in working expenses, and another £3 per mile per week in interest payable on a reduced capital account. Further, it should not be forgotten that a transhipment charge of 9*d*. per ton is allowed (by the Railway Clearing House Regulations) in such cases, and this charge the trader is well able to pay, because it is only a fraction of the saving effected for him by the railway conveying his goods at reduced cost as compared with his former modes of conveyance. The actual cost of transhipment is from 4*d*. to 6*d*. per ton.

It may, in short, be granted that a moderate traffic receipt can pay moderate expenses, and have something left for a remunerative dividend, whilst the same traffic, with heavier outgoings, would have less margin for dividend; and when this is accompanied by the heavy capital account, too often incurred in making branch railways, that which might have been a dividend vanishes altogether.

If the expected traffic and the estimated cost of works will admit of a full-sized railway, by all means let the full gauge be adopted. Or if a proposed railway is likely to form a connecting link between existing railways, it should of course be made on the same gauge as the other lines; but when a branch is not likely ever to be anything but a branch, then it may safely be averred that there are some positive advantages in the break of gauge. For instance, the light railway of narrower gauge cannot be trespassed upon by the heavy engines of an adjoining Company; neither can the trucks belonging to the branch go astray on the main lines: the branch can thus carry on its business with a very moderate supply of rolling stock, which may be not only moderate in quantity, but also in size and weight of vehicles, and therefore much more manageable and economical in working expenses.

It is often argued that a little railway may be all very well for a light traffic, but what is to be done when the traffic increases? To this it may be replied, that one of the greatest blunders has been making railways with too great a regard for the future, and not sufficient consideration of the immediate present. No business can stand against a capital outlay which is out of proportion to the business to be done. It is familiar to all, that the successful manufactories and trading enterprises of this country have been the result of small first outlay, and gradual development of business. In those instances where a new business has been commenced with a large capital and huge buildings, failure and disaster have been only too familiar, even in cases which had all the advantages of modern machinery, a skilled staff, and good business connections. The man of business who begins in a small way keeps his premises and capital fully occupied, and success is the outcome of the demand exceeding the supply. To build a railway many times too big in all its belongings, is simply to postpone the day of its becoming remunerative—a postponement which is for ever adding all sorts of burdens to the capital account.

In some of our colonies full-sized railways have been made with the effect that the beginning of dividend paying is still distant; in others, railways have been made on a reduced scale of 3 feet 6 inch gauge and 45 lbs. per yard; and in many of these, the reports are to the effect that the railways do not pay in themselves, but they contribute to the prosperity of the country, and the public revenue is able to pay the guaranteed interest. Of course, where the country is able to bear it, this is all very well, and no very great harm may result in a growing country from making railways too expensive to be self-supporting. But in the majority of cases it is a first necessity that the railways should pay as such; and in the present work the author has specially in view cases where no Government guarantee is obtainable; and in such instances the size of the railway to be made should be brought down to the level of the probable remuneration forthcoming at an early date.

No.	RAILWAYS.	Ballast.	Sleepers.	Weight of Engines in use.	Maximum Speed.	Maintained Running Speed.	Journey Speed, including Stoppages.	Receipts per Train Mile.	General character of Works.
				tons	miles	miles	miles	pence	
1	Great Britain and Irelan Single lines, 7703; d treble, 96; quadru	erally gravel	10" × 5" × 9'	25 to 50	70	50	45 to 16	67	Average
2	Festiniog	..	..	8 to 16	25	20	16	..	Heavy
3	Isle of Man..	Gravel	9" × 4½" × 6'	17	40	24	15	48	Heavy
4	Ballymena, Cushendall,	..		..	..	..	..	..	..
5	Westleigh	..	3" × 5" × 4'	7	5	..	..	..	Heavy
6	Indian Railways, guaran Single lines, 5233; dou	..	..	..	..	..	..	..	..
7	Indian State Railways ..	..		..	..	..	..	..	..
8	New South Wales ..	broken sand- e and gravel.	10" × 5" × 9'} 9" × 4½" × 8'}	20 to 37	26	..	..	103	..
9	Queensland..	sandstone	..	22 to 30	30	20	..	..	..
10	Adelaide to Glenelg ..	..	..	..	..	..	..	..	Light
11	Victorian Government	stone metal	10" × 5" × 9'	..	..	..	..	105	..
12	Cape Colony, Western D	..	..	..	..	..	..	114	..
13	„ Eastern Dis	..	..	..	..	..	..	..	..
14	Cape Copper Mining Co	..	{ 7" × 3" 9" × 4" }	..	..	..	..	..	..
15	Ceylon		10" × 5" × 9'	..	..	..	..	182	..
16	New Zealand	various	8" × 5" × 7'	11 to 19	25	20	16	60	..
17	Tasmania, Launceston a	Gravel	..	26 to 34	40	25	20	52	..
18	Tasmania, Main Line	blue stone metal.	8" × 4½" × 6'6"	17 to 30	..	23	14	..	..
19	Bolivar—South America	Various.	Wrought iron	8 to 24	..	..	..	..	..
20	Majorca—Balearic Island	Gravel.	9" × 4½" × 6'	20 to 22	34	18	15½	52	..
21	Swedish	..	..	33	..	..	30	..	..
22	„	..	..	..	..	..	23	..	..
23	„	..	..	..	..	..	15	..	..
24	China, Shanghai, and V cluding land and exc penses amounting to per mile)	" broken shingle.	9" × 4" × 5'	9	22	18	15	..	..

CHAPTER II.

THE most costly feature in all railway work is " speed." Both in first cost and in maintenance expenditure is indispensable if speed be demanded. In many new countries the prime necessity lies in being able "to go at all," and speed is quite a secondary consideration. In some cases there may be roads, or perhaps only mule tracks; in such instances the travelling speed for passenger trains may well be limited to 12 miles an hour or even less.

The facilities now at command for enabling rolling stock to turn very sharp curves, render it possible to make surface lines with easy gradients; and if these conditions be followed, and only a light load be imposed on each wheel, a very light permanent way will suffice. It is perhaps scarcely necessary to speak of the desirability of easy gradients, but a few words may here be added on this very important topic.

Formerly gradients of 1 in 300 were striven for, and should still be striven for, especially where high speed is required. Although, in recent years, gradients of 1 in 60 and even 1 in 40 have become frequent, it cannot be too often insisted on, that these severe gradients should only be adopted when they are absolutely unavoidable. An addition of 10 per cent. or more to the length of a railway may be very judicious, if by that means extreme gradients can be avoided. The ruling gradient is in fact one of the prime elements in the determination of the calibre of a railway. As a general rule the gradients of railways fall towards the seaboard, and the paying load is more frequently descending than ascending. Thus it sometimes happens that the measure of the duty of a railway consists in a certain number of empty trucks to be hauled up the ruling gradient; sometimes, however, the whole load has to be taken up the worst gradient. In either case, if there be one or two gradients decidedly exceptional to the general course of the line, the plan may be adopted of dividing the train into two or more portions for the ascent of such gradient. This plan will do very well on railways not overcrowded with traffic; it of course requires the construction of a loop siding at the summit, to enable the engine to return from the front of the train to bring up the remaining portions; this method of course involves some delay, but as soon as such delays become inconvenient, the simple expedient of a pilot or auxiliary engine, allotted to the spot, will enable the train to be taken up whole, and so avoid the delay of dividing it.

There is certainly no simpler way of overcoming the difficulty of gradients. The laws of gravitation are inexorable and unalterable, and no invention of any particular kind of engine can alter the fact, that the duty to be done is to move the given load through a given height and distance, although some inventions may facilitate the performance of the duty in a slight degree.

For going up hill, as for travelling on the level, the simplest form of locomotive is to be preferred; and the engine which best combines steadiness, maximum adhesion, fewness of parts, and greatest efficiency, is undoubtedly the ordinary six-wheel-coupled goods engine with wheels of moderate size.

Heavy engines on many wheels are to be avoided, for although the insistent load on the rails may be distributed by having an increased number of wheels, the lateral blows on the road certainly depend, rather on the whole weight of the engine, than on its load per wheel. It is the mass and bulk of the engine which tears the road to pieces.

Of the various contrivances of recent years for facilitating the ascent of heavy gradients, the most practical is that in use on the Rigi Railway. This is a return to the old Blenkinsop system of a toothed rack, but with this difference, that the toothed rack is applied where it is really needed, and with a perfection of construction leaving little to be desired. Not the least difficulty in very severe gradients is the danger of descent, and the provision for this purpose on the Rigi Railway is most complete. Where gradients steeper than 1 in 25 may be a necessity, and where the trains are likely to be frequent, the experience gained on the Rigi Railway may be turned to good account. The average gradient on that line is 1 in 4 for a distance of about four miles, and the railway has been in successful operation for about seven years. A full account of it is to be found in Transactions of Inst. Civ. Eng. of London, vol. xxxvi., in a paper contributed by Dr. Pole, F.R.S. For such gradients the choice lies between the old-fashioned winding engine, and the Rigi type according to circumstances, the latter having the advantage of safety for passenger traffic, whilst the former may be more economical for mineral traffic.

The first section of the Rigi Railway was opened in May, 1871, and the success attained by it has led Mr. Riggenbach, its designer, to continue his work in other directions. He has already constructed eight other railways on the same principle, with locomotives varying in weight from 10 tons to 30 tons. The latest example is an engine of 30 tons capable of taking four times its own weight up a gradient of 1 in 55 at a speed of 10 miles an hour when working as an ordinary engine, and the same load up 1 in 20 at a speed of 5 miles an hour with the assistance of the rack.*

FIG. 1.

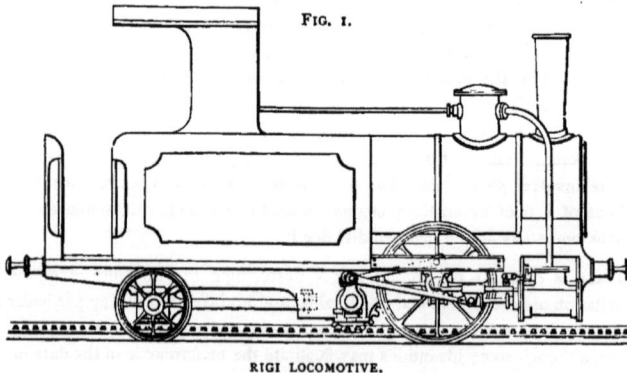

RIGI LOCOMOTIVE.

Fig. 1. Six-ton engine. A favourite type is an engine weighing 6 tons, capable of drawing 40 tons at a speed of 12 miles an hour up 1 in 100; or 24 tons up 1 in 50 at a speed of

* For a fuller account of the Rigi system, see 'Die Drei Rigibahnen und das Zahnrad-System,' published by Messrs. Orell, Füssli, and Co., Zürich, 1877.

9 miles an hour, when working as an ordinary engine; and by means of the rack it can take the same load up 1 in 18 at a speed of 4 miles an hour. The cost of these engines is only about 10 per cent. more than the cost of ordinary engines of the same weight. The rack can be very strongly constructed in wrought iron and steel at a cost of £1 per yard run. This additional cost of course occurs only at the severe gradients.

For gradients from 1 in 20 to 1 in 25 a well-constructed ordinary engine can be depended upon to take a load equal to its own weight behind it, and for a test duty it will do even double that.

To work steep gradients cheaply, the way is to be content to deal with the load a little at a time, and on a passenger line the ruling load should be determined, if possible, by the usual weight of passenger train required to be taken up the steepest gradient without the delay of dividing the train; goods trains being divided as indicated above.

If, however, the goods or mineral traffic is greatly in excess of the passenger requirements, the greatest load to be hauled at one time must be determined by the nature and quantity of traffic to be conveyed during the day. To move a given load per day from one place to another in the most economical manner, the aim should be to divide the work into as many trains per day as may be convenient with a minimum number of engines in steam, remembering always that the maximum load to be hauled *at any one time* up the maximum gradient determines the size and weight of locomotives to be employed, and this again fixes the size and the general conditions of the railway.

A heavy railway with steep gradients should not be made, if, by taking a longer route, easier gradients can be obtained, and consequently a lighter railway can be made to suffice.

There are many heavy railways in different parts of the world, on which the whole traffic of the day is conveyed in one or two heavy trains! Can such railways be said to be properly occupied? Would not a smaller railway with four or six lighter trains be much more manageable, besides giving the advantage of more frequent communication? If a railway be small it will the sooner be fully occupied; and if fully occupied, it will be highly remunerative; and long before the time comes for making additional lines of rail, the financial position of the railway would be such that it would be able to command any required capital on easy terms.

CHAPTER III.

In very many countries abroad the soil is literally teeming with valuable products; sometimes there is great trading wealth in ivory, skins, or other valuables. The difficulty of reaching the coast or a navigable river is usually the barrier to the general prosperity of the population. In such cases the smallest railway quickly executed would be an incalculable boon; and when the products are valuable and of moderate weight, the smaller the railway and the less the expenditure upon it, the better will be the results for all concerned.

With this view the subjoined estimates have been prepared, showing the cost of materials for railway construction and equipment, for lines varying in size and calibre from 2 feet gauge and 12-lb. rails up to 3 feet 6 gauge and 45-lb. rails.

In this task the writer has been very much assisted by the incidental circumstances of his having for many years taken an active interest in the introduction of railways into China. That subject eventually resolved itself into a question of the smallest practicable locomotive,

Fig. 2.

PIONEER LOCOMOTIVE.

which might be sent to that country in the most unobtrusive and inoffensive manner. In the result a little engine was made weighing only 30 cwt. in working order, and capable of travelling 15 miles an hour (sometimes 20 miles), and able to draw two or three trucks of ballast each three times its own weight. The starting of the trucks was of course very severe work, and the use of so small an engine with such trucks was never intended; but it is mentioned as showing what has been done in this direction. This engine is shown in Fig. 2.*

Again, having been called upon to design a means of conveying five or six passengers, at a speed of about 10 miles an hour, up rather steep gradients on one of our colonial railways, the author constructed for this purpose the unpretending but commodious steam carriage shown in Fig. 3. The chief condition was that the machine should be as inexpensive as

FIG. 3.

STEAM CARRIAGE.

possible, and an open carriage was deemed sufficient for the immediate object in view. This machine weighs three to four tons in working order; it has room for six to ten passengers, and travels 20 miles an hour; and, if required, it can take an ordinary carriage or wagon behind it. Although both of the above sizes are on a very small scale, they are not too small for

* NOTE.—The woodcuts of locomotives, briefly referred to in this chapter, are repeated on the pages adjacent to the several estimates, so as to place before the reader, at one glance, a complete view of the appearance, dimensions, and capabilities of the engine, together with an estimate of cost of the other materials and plant to correspond therewith.

commercial purposes. There are certainly many cases in which they may be serviceable either for small railways on private estates or as economical pioneers in distant countries. Estimates relating to both are accordingly subjoined on pages 64 and 66.

FIG. 4.

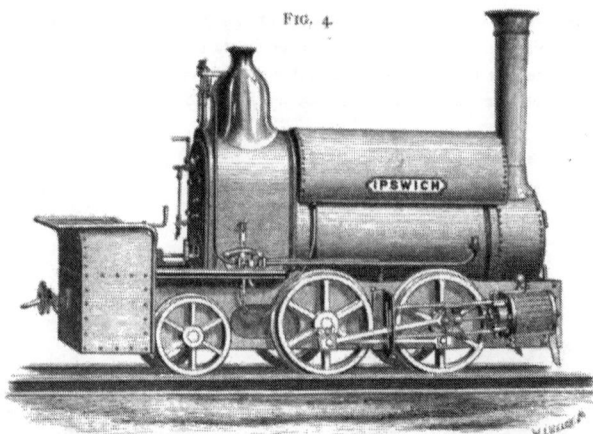

LOCOMOTIVE " IPSWICH."

Fig. 4 gives an illustration of a very serviceable small locomotive engine for general purposes. It weighs 4 tons in working order, and with driving wheels of 2 feet in diameter it runs easily 20 miles an hour with light loads. With the leading wheels coupled it will take a load of 50 or 60 tons at a low speed. For this engine the trucks should be of a size to weigh about 12 cwt. empty and 3 tons when full. Fitted with radial axle-boxes for the uncoupled wheels, the fixed wheel-base is only 3 feet, and the engine will easily go round curves of 1 chain radius.

FIG. 5.

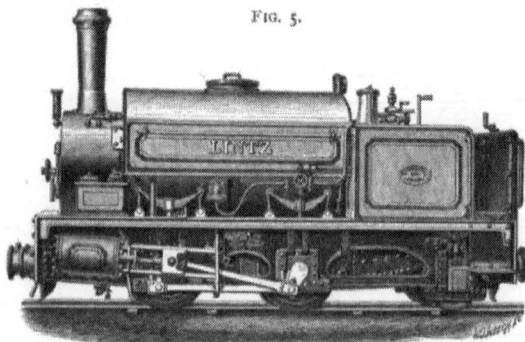

LOCOMOTIVE " LINTZ."

Fig. 5 is a woodcut of a very useful engine built by Messrs. Black, Hawthorn, and Co. It weighs 7 tons in working order. Full particulars of it are given at page 70. It is fitted with a patent arrangement for allowing lateral play to the trailing wheels, without sacrificing steadiness of running. It has done good service on mineral lines, and it is well adapted for general purposes where the loads and speeds are moderate. The same engine is sometimes made with four wheels only and a total wheel-base of 4 feet.

FIG. 6.

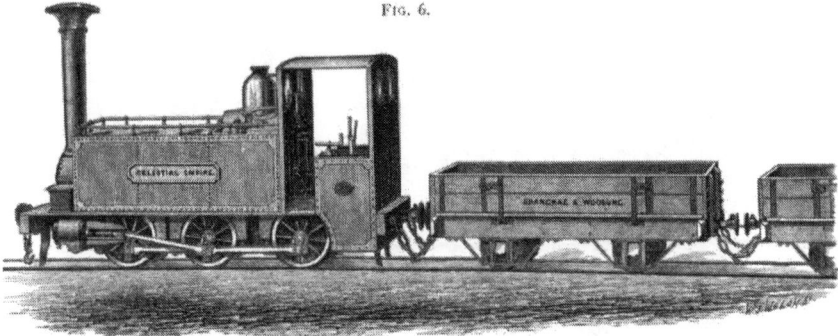

LOCOMOTIVE "CELESTIAL EMPIRE."

Fig. 6 is from a photograph of the engines "Celestial Empire" and "Flowery Land," designed by the author, and built by Messrs. Ransomes and Rapier, for the opening of the first railway in China. These engines weigh 9 tons in working order, and have wheels 2 feet 3 inches diameter. The quality of the water being bad, and the supply uncertain, they were made with very large side tanks, so as to carry more than sufficient for a double trip. Full particulars

FIG. 7.

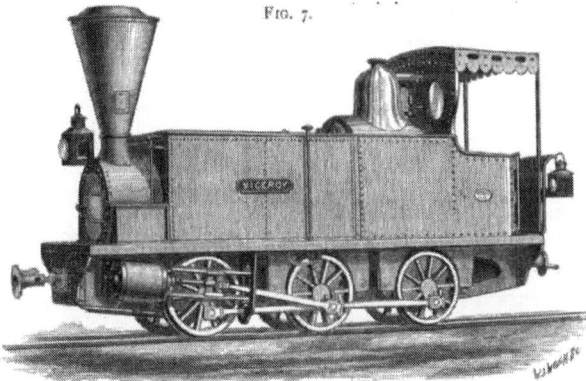

LOCOMOTIVE "VICEROY."

as to their dimensions will be found on page 72. Here it may be added that these engines have travelled at the rate of 25 miles an hour, and easily maintain a speed with passenger trains of 18 miles an hour, on a line with occasional gradients of 1 in 132.

Fig. 7 is a woodcut of the engine " Viceroy," since designed and built for the same line, as a sequel to, and a size larger than, the former engines. The weight of this engine is 12 tons, and the wheels are 2 feet 6 inches diameter, thus giving a step forward in the direction of a little more speed and a little more power, without interfering with the general character of the railway, which was of necessity of a light pioneer type.

FIG. 8.

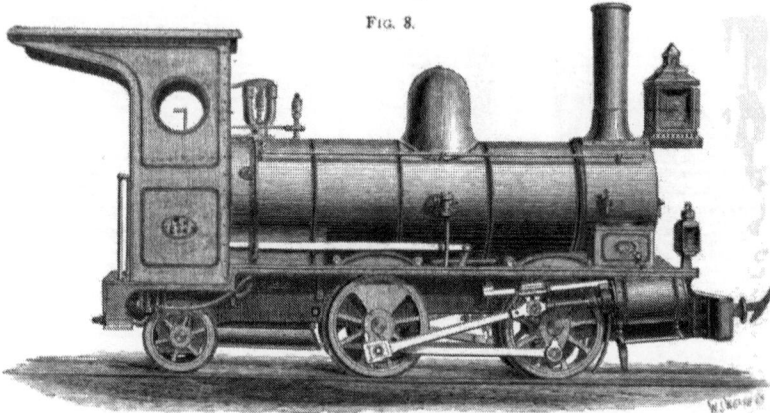

LOCOMOTIVE IN USE ON INDIAN STATE RAILWAYS.

Fig. 8 is a woodcut of an engine weighing 15 tons in working order, such as are now used on the Indian State railways, of metre gauge and 40-lb. rails. The tender is not shown in the woodcut.

These engines have four coupled wheels 3 feet 6 inches diameter, and two trailing wheels 2 feet 9 inches diameter. They are well adapted for working mixed trains of passengers and

FIG. 9.

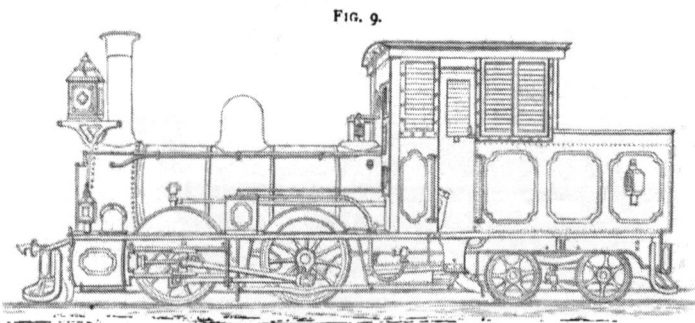

LOCOMOTIVE IN USE ON INDIAN STATE RAILWAYS.

goods. They maintain a journey speed of 15 miles an hour with average loads of 150 tons where the ruling gradient does not exceed 1 in 150; and being fitted with tenders they carry fuel and water sufficient for 40 miles under ordinary circumstances. Sometimes the tenders are fitted with sockets to receive stanchions suitable for carrying large supplies of wood fuel. The same size of engine is sometimes fitted with a small tender, mounted on the same frame as the engine; a four-wheeled Adam's bogie being used instead of a single pair of trailing wheels; as shown in Fig. 9. In this case the fixed wheel-base is only 5 feet, although the total wheel-base is 17 feet 6 inches. Large numbers of these engines (of both types) have been built by Messrs. Neilson and Co., of Glasgow, by Messrs. Dubs, of Glasgow, and by Messrs. Nasmyth, Wilson, and Co., of Manchester.

Fig. 10 is a woodcut of an engine in general use on the railways of New Zealand.[*] Its weight in working order is 18 tons; it has six wheels 3 feet 6 inches diameter, and all coupled. This is a highly efficient type of engine for steep gradients, the whole weight being utilized for gaining maximum adhesion. Further particulars are given in estimate No. 8.

FIG. 10.

LOCOMOTIVE IN USE ON NEW ZEALAND RAILWAYS.

To go beyond these sizes would be to enter the region of full-sized railways and engines, the consideration of which is foreign to the present purpose. The object the author has in view is essentially cheap, small, slow-speed railways.

Here it may be proper to remark, that care must be taken to guard against the error of making small railways and then expecting them to do the duty of heavy ones. For example, to make a railway of 3 feet gauge and 10-ton engines, and then attempt to travel at 25 miles an hour, or to work a goods traffic with trucks which weigh 3 tons empty, is to mistake the whole nature of the subject.

* New Zealand has made more rapid progress in the development of light narrow-gauge railways than any other colony. At the end of June 1876 there were 860 miles of railway open for traffic, and 367 miles in progress.

Gauge, weight of engines, diameter of driving wheels, and weight of rails must be all duly proportioned to the weights to be conveyed, and to the speed at which they are to be drawn ; and, having fixed the dimensions of the means, the duty originally marked out should be rigidly adhered to.

It has already been remarked that a gauge of 4 feet 8½ inches, with 80-lb. rails and 50-ton engines, with 7-feet wheels, may be considered the minimum conditions of a line requiring a maintained speed of 50 miles an hour, and a maximum speed of 60 to 65 miles.

Similarly a gauge of one metre, with 40-lb. rails and 15-ton engines, with 3 feet 6 inch wheels may be considered equal to a maintained running speed of 22 to 24 miles an hour, and a maximum speed of 28 or 30 miles with light loads.

As a rule no practical advantage is to be gained by making the gauge less than 2 feet 6 inches, and, all things considered, 3 feet gauge is better, but in new countries much may be done on a 2 feet 6 inch gauge, with 25-lb. rails and 9-ton engines, having driving wheels 2 feet 3 inches to 2 feet 6 inches diameter, if the speed be limited to a maximum rate of 20 miles an hour, and a maintained speed of 16.

When it is borne in mind that these are the speeds at which branch line traffic in this country is chiefly worked, it becomes strikingly apparent what useless waste has been going on, in making branch railways on such a scale, that many of them cannot pay for several generations.

In all cases, however, where there are no special circumstances suggesting a gauge of 2 feet 6 inches or less, it would be advisable to adopt 3 feet as a standard, because it admits of more commodious vehicles than the narrower gauge does, and if in the course of time a higher speed be demanded it can be obtained by gradually increasing the power of the engines and the weight of the rails as they are from time to time renewed. For example, a line of 3 feet gauge and 25-lb. rails might begin with a maximum speed of 20 miles, and a working speed with light loads of 15 miles, and a goods train speed of 10 miles an hour. And by gradually increasing the rails to 50 lbs. per yard and improving the permanent way generally, the maximum speed might be correspondingly increased up to 35 miles an hour. These improvements can all be effected gradually, and as the line might be able to afford them ; and they might be made first in those parts of the line where such additional outlay would be most likely to be remunerative. In cases where they are likely to occur at an early date, and particularly if speed is likely to be soon required, it would be best to adopt a gauge of 3 feet 6 inches. This gauge is not too wide for a light line and light rolling stock, and it is wide enough for considerable speed. In saying this, however, the author would again add the word of warning that railways have often been made too big.

It has been already hinted that the gauge of 2 feet 6 inches, and the general calibre of the Shangai and Woosung Line were adopted for reasons of policy and economy, and not in any way as being most suitable for that populous district. This opportunity has, however, abundantly proved the capabilities of such lines for good practical work in new countries.

The 9-ton engines of the " Celestial Empire " class have been constantly taking 200 to 250 passengers at a time, at a speed of 18 miles an hour. In a large majority of instances such a number of passengers to be conveyed at one time is quite the outside of any immediate requirements in new countries, and to make any larger provision is simply *to try to prevent* the railway paying.

But it is sometimes said that railways should be made with due regard to future increase of traffic. It has already been pointed out that it would have been better if more attention had been given to immediate wants, and it would be far better to let the prime question be " How soon is this proposed railway likely to be fully occupied ?"

As to what constitutes full occupation for a railway, vide Transactions Inst. C.E., vol. xxxviii. p. 177.

In practice it is found that one line of metals can well accommodate 200 trains per day of 16 hours, if the trains are all travelling in the same direction and at the same speed, or 100 trains per day if they travel at different speeds.

For single lines with trains travelling alternately in opposite directions, the number of trains per day depends chiefly on the distance apart of the passing places.

With passing places 10 miles apart, and a journey speed of 15 miles an hour, it is possible to have 8 trains each way per day of 16 hours. Or if the trains be run two consecutively in the same direction, with a quarter of an hour interval between them, 12 trains per day each way can be accommodated.

If the passing places be only 6 or 7 miles apart, the practicable number of trains would be 50 per cent. more than the above.

CHAPTER IV.

RAILWAY expenditure may be estimated on a sound basis, as follows :—

Let the probable value in £ sterling of the traffic, per mile per week, during the first year or two of the working of the proposed railway be carefully estimated. It is tolerably certain that at least half of the gross receipts will be absorbed in working expenses, so let the estimated gross receipts be divided by two, to give some idea of the net revenue per mile per week in pounds, and this may be taken as a precise index of the number of thousands of pounds per mile to which the expenditure should be limited, in order to ensure 5 per cent. dividend.

For example :—For a proposed railway of 20 miles let the quantity of goods likely to offer be 100 tons per day (50 tons each way), at 4s. per ton = £20 ; and the number of passengers 200 (100 each way), at an average of 2s. each = £20; total traffic £40 per day, or = £240 per week = £12 per mile per week.

For such a railway in distant countries the expenses may be estimated as follows :—

	Per Week.
1 engineer	£16
1 general foreman	6
3 mechanics (to act also as drivers)	15
5 station masters and guards	15
Platelayers and porters	28
Fuel and oil	12
Materials for repairs (for the first year or so)*	20
Stores, stationery, &c.	8
	£120 =

£6 per mile per week.

This deducted from £12 per mile per week gross receipts, would leave £6 per mile per week net receipts, and would give 5 per cent. dividend on an expenditure of £6000 per mile.

But the cost need not exceed £4000 per mile. For earthworks, ballasting, fencing, bridges, &c., &c., so much depends on local circumstances that it is difficult to give figures. The width of formation in general use on high-speed lines is for the single lines four times the

---

* A sufficient supply of materials should be taken on capital account to cover the first year's maintenance, and to ensure in fact the full completion of the line in thorough working order.

gauge. For light railways, with light loads moving at low speeds, three times the gauge may well suffice. Thus, for a light railway of 3 feet gauge a width of 9 feet at formation level will do, and this, in the case of a surface line requiring as an average 15 inches height of table and slopes of 2 to 1, would involve 3300 cubic yards per mile, including an allowance for side ditches. The cost per cubic yard depends, of course, on the nature of the soil, and the price of labour on the spot, but the cost will seldom be found to exceed 1*s.* 6*d.* per cube yard, and this would give £250 per mile.

The cost would be somewhat as follows :—

	Per Mile.
Land and expenses connected therewith	£200
Bridges and culverts (apart from costly viaducts)	180
Fencing	180
Earthworks on a surface line	250
Ballasting, rail laying and lifting	200
Five stations, at a total cost of £2000	100
One workshop, £1500	75
Materials and rolling stock from England (Estimate No. 5)	1735
Freight and insurance on same	400
Unloading, erecting, and fixing	100
Supervision	200
Contingencies	380
	£4000

NOTE.—This cost may be reduced by £300 per mile if sleepers are to be had on the spot.

The general conditions of such a railway would be as follows :—

1. Engines of similar type to those made by Messrs. Ransomes and Rapier for the first railway in China (Shanghai and Woosung).

2. Engines weighing in working order 9 tons, and having six wheels, 2 feet 3 inches diameter, all coupled.

3. Cylinders, 8 inches diameter by 12 inches stroke. Heating surface, 210 square feet. Grate area, 4 square feet.

4. Greatest weight on any wheel, 30 cwt. Rails, 25 lbs. per yard. Gauge, 3 feet.

5. Wheel base of engines, 8 feet 6 inches. Wheel base of carriages, 9 feet.

6. Sharpest curve practicable at 10 miles an hour, 5 chains. Sharpest curves practicable at shunting speeds, 3 chains.

7. Maintained speed, 18 miles per hour, with eight carriages and 180 passengers, on a fairly level line.

8. Maintained speed, 10 miles an hour, with a goods train of 120 tons gross, on a fairly level line.

9. Maintained speed, 5 miles an hour, with a goods train of 60 tons gross, up a gradient of 1 in 100.

10. Maintained speed, 3 miles an hour, with a goods train of 30 tons gross, up a gradient of 1 in 35.

11. A net earning of £6 per mile per week would pay 7½ per cent. on an outlay of £4000.

12. It has been already shown that such net revenue would be the result of 50 tons of goods and 100 passengers per day each way.

13. As the single line of way, with suitable passing places, would accommodate ten or twelve times the above, there need be no fear as to a railway on this scale being adequate for an increased traffic.

It is thus shown that whilst a gross traffic of less than £12 per mile per week cannot support a full-sized railway, a smaller railway costing £4000 or £5000 a mile can live on that very light traffic (and under favourable circumstances on even less); and at the same time if the traffic should increase to £50 a mile a week the line can still carry it; and, by reason of its first small outlay, it would be placed in such a position financially that a double line could easily be laid when required. By the gradual introduction of heavier rails, at each time of renewal, the line might in course of time approximate to the conditions of a full-sized railway; but this would be after the moderate first expenditure had developed the traffic able to support the heavier, stronger, faster, and more expensive railway.

In the above case the expense of a professional engineer and skilled staff at high rates of pay, tell heavily in the working cost of a short railway, and a longer line would show to better advantage. Under very favourable circumstances the above cost might be kept down to £3000 per mile, by adopting Estimate No. 4 and cutting down everything in proportion.

In making similar railways at home, the additional cost of land and parliamentary expenses would be compensated by the saving of freight and insurance, and by some economy in other items.

CHAPTER V.

It may be well now to add an estimate for a light railway for a sugar estate, or for mining, or other purposes where the railway forms a part of an enterprise, and does not require separate supervision.

In such a case the expenses may be set down as follows, for a short line of say ten miles :

Weekly expenses—1 leading mechanic } both drivers {	£6
1 second mechanic } {	5
1 foreman platelayer	5
Platelayers and gatemen	10
Fuel and oil	6
Materials for repair (for the first year or so) ..	5
Stores, &c.	3
	£40

For a surface line of 2 feet gauge, in a level district, a formation of 6 feet wide and 1 foot high would suffice, or about 1200 cubic yards per mile, which at 2s. per yard = £120 per mile.

The cost of such a line need not exceed £2000 per mile, as follows :—

APPROXIMATE ESTIMATE.

Land way leaves, &c., per mile	£150
Bridges, &c...	180
Earthworks	120
Ballasting, rail laying and lifting	200
Materials from England, Estimate No. 3	800
Freight and insurance	125
Unloading, erecting, and fixing	100
Supervision during construction	150
Contingencies	175
Per mile ..	£2000

This expenditure would require £20 per week for interest of capital, and if £40 per week for expenses be added, a gross sum of £60 per week would be required to render the outlay remunerative.

To earn £60 per week, or £10 per day, 80 tons, at 2s. 6d. per ton, must be obtained, or any less quantity must be charged at proportionately higher rates. Any greater quantity could

VIEW OF SMALL RAILWAY AT WATERSIDE WORKS, IPSWICH.

easily be conveyed by the railway, up to say 400 tons per day, on a single line of rails with a passing place at the middle of the journey.

Railways still smaller than the above have done excellent service, for what may be called private use as distinct from public traffic. At Woolwich Arsenal and at the Crewe Works of the London and North-Western Railway, lines of 18-inch gauge are laid throughout the works for the conveyance of material of all kinds with the aid of little locomotives built for the purpose. These lines have been in use for some years, and are now indispensable. Their very narrow gauge, and the short wheel-base of the vehicles in use, enable curves of 20 feet radius to be used, and thus the little railway penetrates into every corner of the premises.

A still more striking instance occurs at the Waterside Works, Ipswich, where may be seen a railway of only 14 inches gauge. A woodcut from a photograph is shown in Fig. 11. This railway is found to be so convenient for a particular purpose, that there is nearly a quarter of a mile of it, running in all directions, on a space scarcely larger than half an acre. The line is complete with points, &c., and has turntables, about 3 feet in diameter. Its trucks weigh 2 cwt. empty and 16 cwt. laden ; the gross load being no less than eight times the dead weight, a proportion which is not attainable in large wagons, nor in wagons of much broader gauge. Much of this railway has been at work twenty years, and the cost of repairs is inappreciable ; the speed is a walking pace, but might easily be 5 or 6 miles an hour. The rails are made of cast iron (the two side rails and the cross-ties being cast in one piece), and they are mounted on longitudinal timbers of light scantling, as shown in the woodcut. This construction answers very well, when the tram-plates can be made in the locality, and without involving great cost for carriage to their destination. But for similar lines for general use, it is decidedly better to use wrought iron or steel rails, mounted on transverse wooden sleepers, as shown in Fig. 12, or on wrought-iron sleepers, as shown in Fig. 13. The rails and sleepers are prepared with a view of being put together as speedily as possible and formed into lengths of road at their destination. When once put together, they need not be taken apart again, as each 12-foot length weighs only 2 to 3 cwt., and can be readily carried by two men. This kind of road is really portable, and will be found to be well adapted for any purposes, for which the rails are required to be moved from one place to another, from time to time ; as in collecting sugarcane or other heavy crops, or in moving building materials or minerals. The several lengths may be joined together, either by fish-plates or by oak cleats nailed on the joint sleepers, as shown in Fig. 12, so as to make sockets to receive the ends of the rails of the adjacent length.

Approximate cost per yard of light roads in prepared lengths, including rails, sleepers, and fastenings :—

Weight of Rails.	Gauge, 14 in. to 2 ft.	Gauge, 2 ft. 6 in. to 3 ft.
12 lbs. per yard.	4s. 6d. per yard.	5s. 6d. per yard.
15 ,, ,,	5s. 0d. ,,	6s. 0d. ,,
20 ,, ,,	5s. 6d. ,,	6s. 6d. ,,
25 ,, ,,	6s. 0d. ,,	7s. 0d. ,,

The smaller sizes may be used for temporary purposes of all kinds with very great economy. In moving building materials the smallest size enables bricks, &c., to be moved at about half

FIG. 12.

12-FT. LENGTH OF PORTABLE RAILWAY WITH WOODEN SLEEPERS.

the cost of wheelbarrow work. In short, to move any considerable tonnage from one place to another, a few lengths of this temporary road afford the readiest and cheapest means.

FIG. 13.

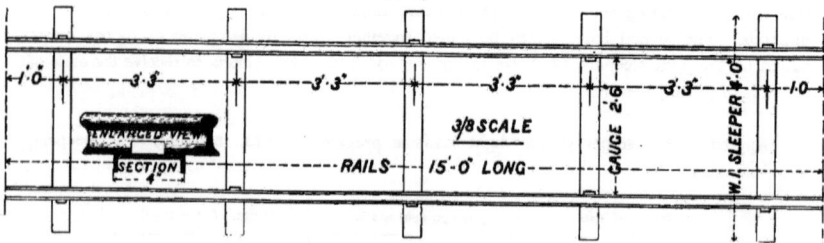

PORTABLE RAILWAY WITH IRON SLEEPERS.

These portable railways will also be found to be very useful as adjuncts to the small railway sketched on page 28. That with wooden sleepers is best for soft ground, and should be preferred wherever wooden sleepers are practicable.

CHAPTER VI.

In thus advocating light and cheap railways the author does not propose to sacrifice anything in quality. Real economy is only to be had in being content with smaller size and less speed; and, consequently, lighter works and plant. The precise scale for any particular line must be determined by the local considerations, but in all cases the smallest well-constructed railway is certain to pay the earliest and the best. Nor should the fact be lost sight of that a little railway, being as free as possible from the incubus of heavy expenditure and accumulated arrears of interest, will the sooner be able to afford to become a double line, and a double line will accommodate from four to six times as much traffic as a single line. Again, if, in course of time, a double line should be insufficient for the traffic, additional lines of rail could be laid, and thus facilities would be afforded for separating the fast from the slow traffic—a desideratum now fully recognized by our English Railway Companies.* Seeing that this has become a pressing necessity after only thirty years of work, it should certainly be kept well in view in laying out main lines of railway in other countries.

It may be well to make now some brief reference to the several particulars of construction, plant, and rolling stock, which should be kept in view.

BALLAST.

One of the most important parts of a railway is a bed of firm, and easily drained, material, on which to lay the sleepers. To obtain a plentiful supply of such material is of so much importance in the permanent maintenance of a railway, that it should be kept in view from the very outset; and it will be found well worth while to make even considerable deviation from the true line of road, if, by so doing, better supplies of natural ballast can be obtained.

The bed for the sleepers must be raised above the adjacent formation, and provision should be made for its constant drainage. In cuttings, side ditches must be made for this purpose; the various gradients usually give all that is required in the way of fall and eventual outlet. Water tables are commonly made between two sleepers at intervals of 10 feet along the length of the line; generally two such passages are given in the length of a rail. The practice of making one of these tables or passages at the joints of the rails is very frequent; the reason being that the partial removal of the ballast affords facility of access to the fish bolts for tightening up, &c.; but it would seem to be better to make the water tables in the second

---

* Vide Transactions Inst. C.E., vol. xxxviii. p. 179.

spaces beyond the fish joint, and so have the sleepers adjacent to the joints and the central sleeper better supported ; to have the rails and sleepers well secured at the centre of the length of each rail is very important in order that the expansion or contraction of the rails may be evenly accommodated.

The kind of ballast most used in England is gravel. When this is not obtainable, burnt clay, resembling small pieces of broken brick, is used, and in some cases slag from ironworks, broken stone, chalk, &c. In colliery districts cinders and even coals are used.

The quantity of ballast used on different railways varies very greatly. It depends very much on the quantity and quality obtainable. On some of the English railways the sleepers are entirely " boxed up " with ballast ; that is, the ballast is laid on so liberally as to entirely enclose the ends of the sleepers, and sometimes it is also laid over the sleepers and up to the neck of the rail. This partial covering of the rail protects it somewhat from the excessive heat of the sun in hot weather.

The quantity (for each pair of metals) varies from three cubic yards per yard run on a high speed railway, to none at all on a contractor's temporary road ; the sleepers in the latter case being adjusted by packing in the surface soil. Such work is, of course, only temporary.

It is a common practice to lay the rails and sleepers, and make them subserve the purpose of conveying the ballast before they are themselves ballasted.

For cheap railways, if ballast has to be brought from a distance at great expense, only so much should be used at first, as may be absolutely necessary ; to get the railway practicably at work, should be the first consideration ; improvements of the ballasting can be made rapidly and cheaply with the aid of the line and plant, for then the ballast can be brought from a greater distance, and a selection can be made as to quality ; whereas in the first instance the nearest obtainable may suffice. In all cases the line should be well ballasted before engines are allowed to run at any speed much above a walking pace.

SLEEPERS.

The practice of using as many sleepers as possible, has great advantage in cases where ballast is expensive or difficult to obtain ; the increased bearing surface, thereby obtained, being of much benefit. Indeed, it may be stated, that the cheapest and best road for light work is certainly to be made with rails of Vignoles type, and with the sleepers as near together as possible, consistent with having sufficient space between them for access for beating up. For full-sized rails, sleepers are commonly laid 3 feet centre to centre, but for small rails, this may with great advantage be reduced to 2 feet ; and if the sleepers be 8 inches broad, there is still 16 inches space between the edges of the sleepers, wide enough for purposes of repair, and at the same time close enough to ensure full supporting and tying of the road.

The different sizes of sleepers suggested for different sizes of railways, are given in the estimates in the succeeding pages.

Sleepers of rectangular section, are much to be preferred to half-round. Half-round sleepers, with rail seats cut with the adze, not only give a very poor bearing for the rail, but for a road of equal strength, require a considerable addition to the number of sleepers used, whereas the sawn rectangular sleeper gives a maximum of bearing, and a minimum of distance between the supporting edges.

Although hand-adzing is objectionable, it is a very good plan when machinery is available to cut recesses about $\frac{1}{2}$ inch deep to receive the rails; machines are now made to cut both the recesses at once, and thus provide not only for the cant of the rails, but also for the gauging of the road in laying it, and the shoulders also give some permanent lateral support to the rails. Recesses cut by hand are of no use for these purposes.

Half-round sleepers laid on the round side, or V section sleepers, should never be adopted, as they are continually thrusting the ballast away from them.

Creosoting adds enormously to the durability of sleepers, and should be adopted whenever practicable. There is often, however, very great difficulty in getting ships to take creosoted sleepers, owing to the injury caused to other cargo by the proximity of creosote. This detrimental influence stays by the ship for a voyage or two, after the offending sleepers have been discharged.

Sleepers sent from this country are almost always of Baltic fir. In many countries native woods are to be had of excellent quality; and to suit such cases, estimates are subjoined of the necessary plant for cutting out, seating, and boring sleepers. *Vide* Appendix I. When the sleepers are of hard wood, it is necessary to bore holes, although the spikes may be small. A square spike will find its way very well into a round hole, bored one-sixteenth to one-eighth of an inch less in diameter, according to the hardness of the wood.

In some countries wood sleepers perish very rapidly, owing to the ravages of insects, and for such cases many kinds of iron sleepers have been devised. In Egypt and in India the well-known pot sleeper has been largely used for many years, as also Mr. Livesey's wrought-iron sleepers, and Mr. Batho's, made of steel. Whilst possessing the merit of greater durability, the iron sleepers are, of course, much more costly than wood. For estimating purposes the cost of iron sleepers may be taken as double that of wooden sleepers per mile.

E

RAILS.

Rails are now made to a great extent of Bessemer steel; and at present, at prices which tread very closely on the heels of the iron rails. In course of time it seems probable that steel rails will supersede iron altogether. This, however, cannot yet be said to be the case, especially for small sizes. For small rails, the price per ton is, of course, always higher than for larger ones, and this difference is as yet greater in steel than in iron. It is also to be borne in mind that small rails are generally used where the traffic is light, and for railways of this class saving in first cost is of prime importance. If the anticipated traffic is so light that iron rails will probably last ten years, then there is certainly no advantage in adopting steel at a higher cost.

FIG. 14.

12 LBS.

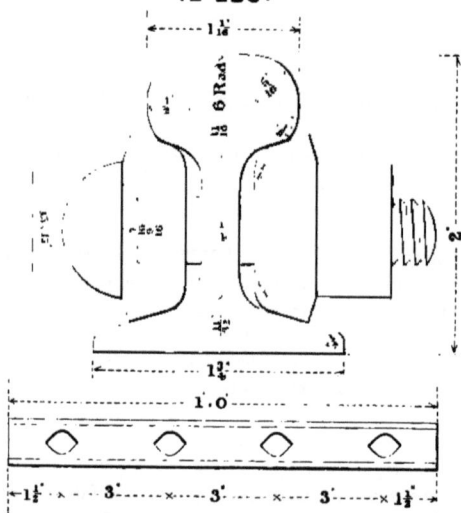

SECTION OF RAILS.

It has been suggested that steel rails may be made lighter for equal strength, and so effect a saving in freight as well as in weight of material. This will probably be the case soon, but, up to the present time, the uncertainty as to strength of steel rails is very great. A hole made in the flange of an iron rail reduces its strength in a degree corresponding very nearly with the reduction of quantity of material, but in steel rails the strength is reduced 80 per cent. by a notch or hole equal to only 15 per cent. of the section of material. The improvements in the manufacture of Bessemer steel may well induce the hope that in a few years these difficulties and anomalies may be removed. For the present, for light railways, iron, as a rule, is to be preferred. The rails should be made from a pile of good hammered iron, arranged so as to give a hard top and a fibrous flange. The varieties of section seem to be almost infinite. A few years ago, Mr. Sandberg ascertained by careful experiment, the proportions of section giving the greatest strength for a given weight of material, and sketches of his sections for different weights of rails are, with his permission, appended. Figs. 14 to 21. It will be observed that the seat for the fish plate is made of the same angle under the shoulder and at the foot, and at an angle well adapted for tightening up. No desire for economy should ever lead to discarding

FIG. 15.

16 LBS.

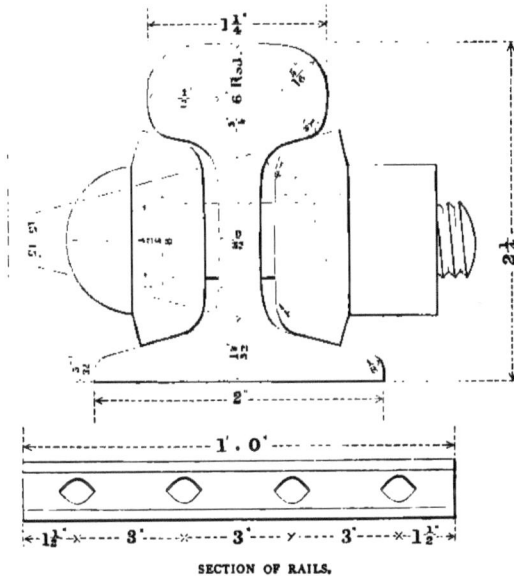

SECTION OF RAILS.

the fish plates. The smaller the rail the greater its need of such an efficient adjunct; it is far better to reduce the weight of the rail than to *save* the fish plates. The same may be said as to the use of fang bolts or coach screws at or near the ends of the rails.

Two fang bolts or coach screws in the sleeper on each side of the joint, and two in the middle of the length of the rail, will be found to be money well spent. Dog spikes will suffice for the other sleepers. The fang bolts should not pass through the flanges of very small rails,*

---

* It is to be noted that in laying down very small rails the fish plates should not be screwed up tight until an engine commences running. If screwed up when there is nothing running on the road the fish plates rust, and grip the rails so tightly, that they cannot slide when expansion takes place, and the road becomes very much distorted by the bending and bulging of the rails.

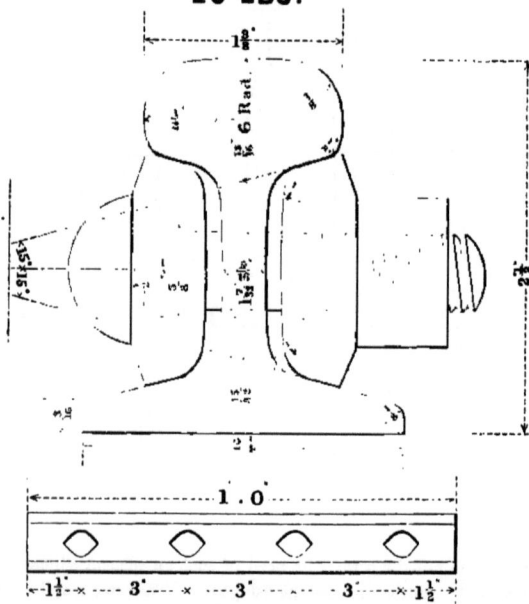

Fig. 16.

20 LBS.

SECTION OF RAILS.

Fig. 17.

25 LBS.

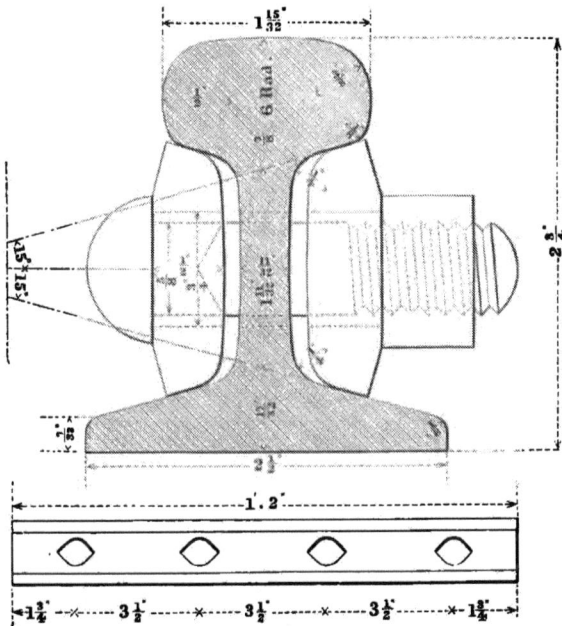

SECTION OF RAILS.

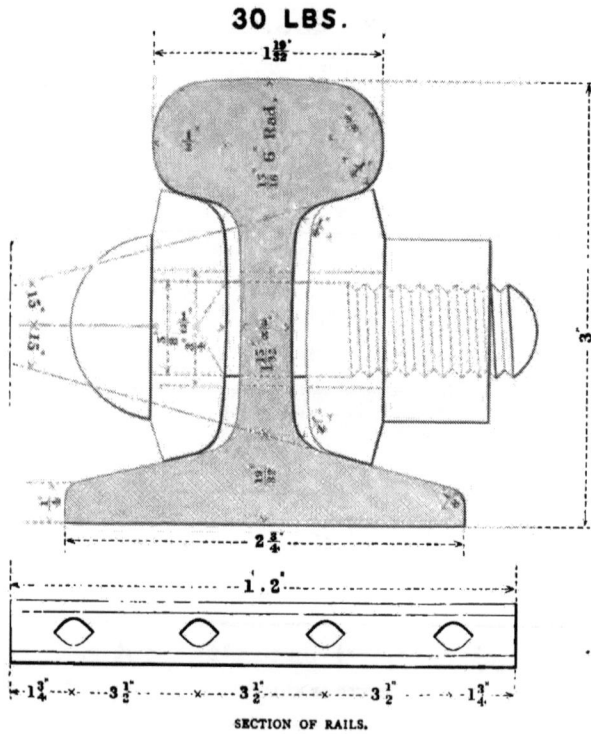

Fig. 18.

30 LBS.

SECTION OF RAILS.

Fig. 19.

35 LBS.

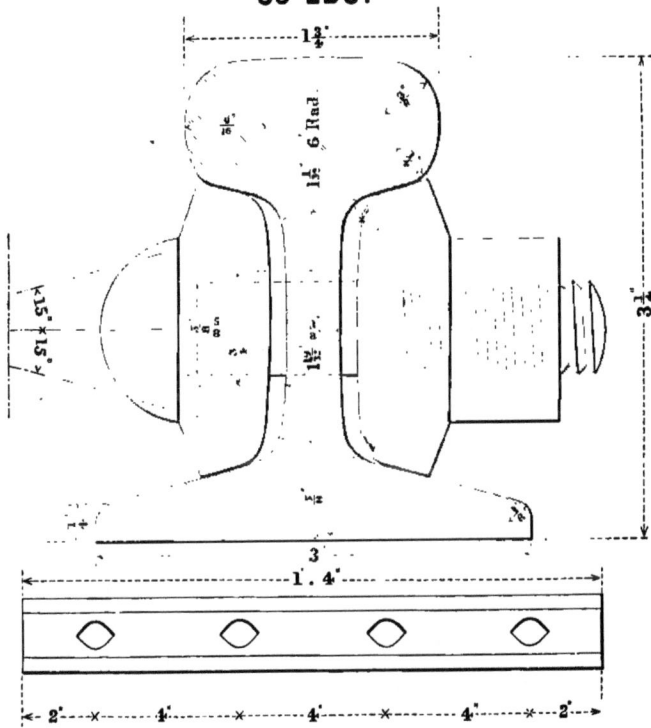

SECTION OF RAILS.

FIG. 20.

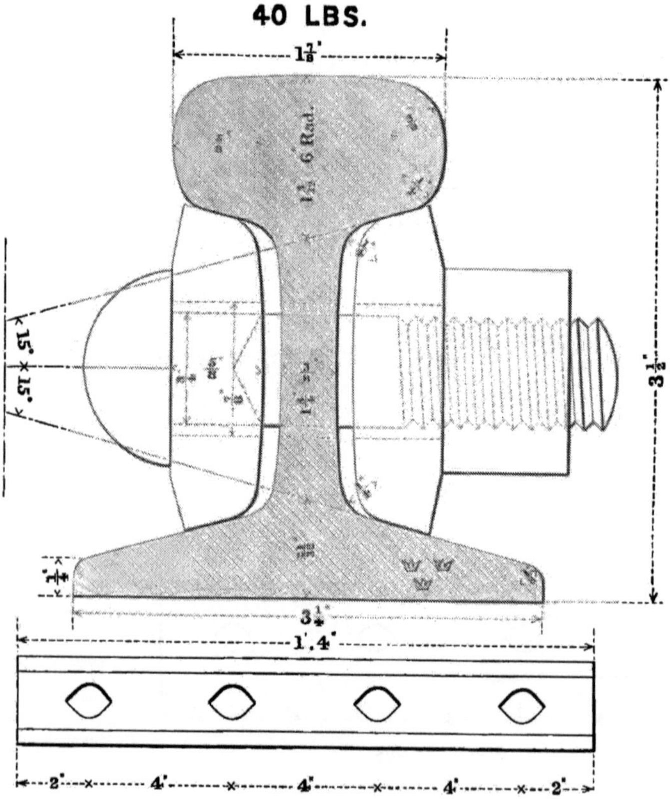

40 LBS.

SECTION OF RAILS.

FIG. 21.

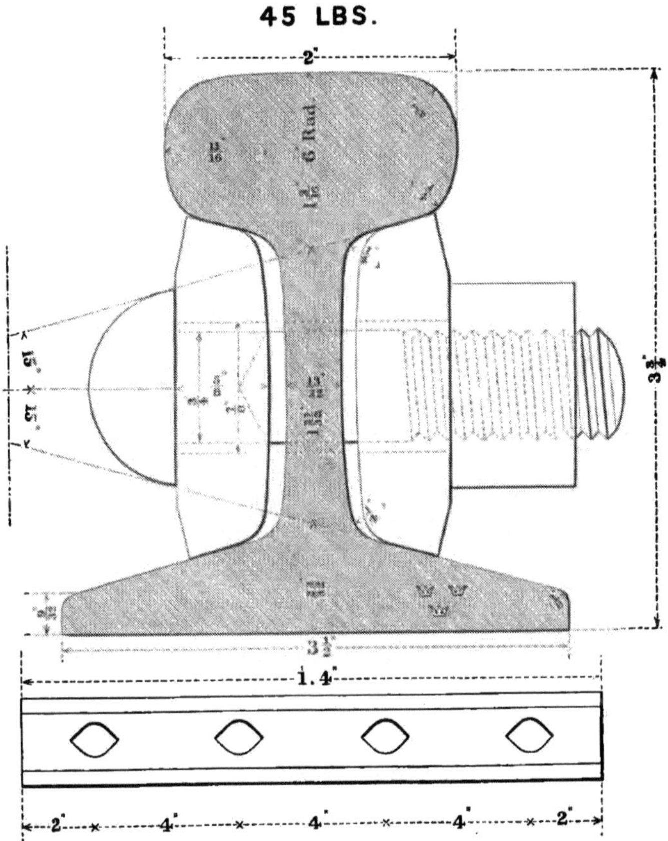

45 LBS.

SECTION OF RAILS.

but should touch the flange, and a clip washer should be used, bearing partly on the sleeper and partly on the flange of the rail. A still further improvement may be made by the addition of seating plates, as shown in Fig. 22, by means of which it will be seen a maximum of support

FIG. 22.

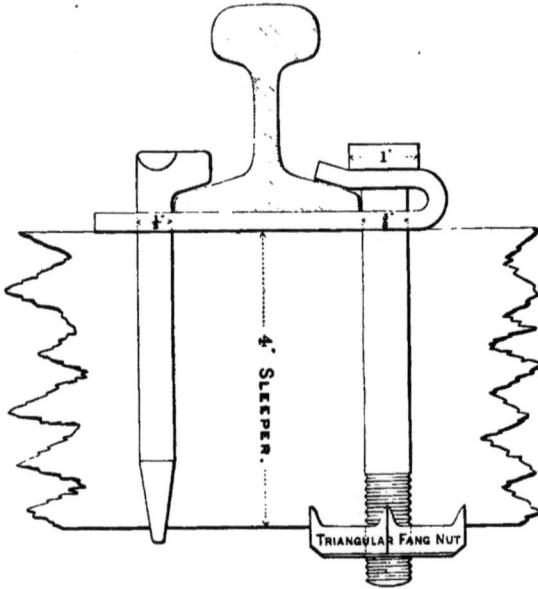

RAIL SOLE-PLATES AND FASTENINGS. SCALE, HALF SIZE.

is derived from both bolts as against lateral strain. These seating plates may be used with great advantage at the middle of each rail, and at the sleepers adjacent to the joints.

POINTS AND CROSSINGS.

Next after the road itself the most important fitting is the switch and crossing. Frequent opportunities of clearing the main road are, in all cases necessary, but especially so on single lines of way. The length of switches and the angle of the crossings must be proportioned to the radius of turn out and to the gauge; a narrower gauge requires a finer angle of crossing than a broader gauge does for the same radius, but quicker curves are used with narrow-gauge railways. Angles of 1 in 6, 1 in 7, and 1 in 9 will be found generally serviceable ; the switch tongues will do very well from 6 feet for 15-lb. rails, to 9 feet for 30-lb. rails, and 10 feet for 40-lb. rails. The number required will vary from five to twenty sets for each terminal, and from two to ten sets for each intermediate station, according to the character of the line and the number of sidings required at each station or passing place. In all cases the rails of the points and crossings should be planed and truly fitted; the chairs for the stock rails are best made of cast iron; sometimes wrought-iron plates are used instead of chairs, but these give scarcely sufficient support. Wrought-iron chairs with a jaw to support the rail on the outside are also made, but they are very extravagant in cost. All switches should be fitted with levers and weights, those on the main line being made so as to " recover " always to the same side; those on sidings may with advantage be made so that they can be set either way at the will of the pointsman, without sacrificing their self-acting character.

The crossings may be either chilled crossings made in one piece or may be built up of rails. The chilled crossing has the advantage in simplicity and great durability, while the rail crossing is more elastic and pleasant to ride over. The rails should be planed to form the V of the crossing and bolted or riveted together; the wing rails should be secured to the V by long bolts passing through from side to side, with distance blocks between the rails wherever the bolts pass through. Crossings so made may sit direct on the sleepers, but it is better to mount them on wrought-iron plates about ½ inch thick.

The rails for points and crossings should, if possible, be of steel; the guard rails for crossings may be of iron.

Before leaving the subject of points and crossings, mention should be made of the modern system of double slip points, now very generally adopted for the facility which they afford for

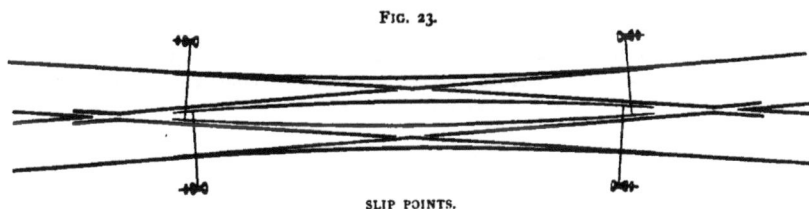

FIG. 23.

SLIP POINTS.

breaking up goods trains quickly. Reference to Fig. 23 will show that access can be had from one road to another with great freedom. In ordering materials for railways abroad a few sets of this apparatus should always be included.

TURNTABLES.

Turntables are necessary where tender engines are used; engines should only be allowed to run with the tender first in an emergency. With tank engines the use of turntables is not absolutely necessary, but it is very desirable, as running the forward way is better for the engines, and the men can keep a better look-out than when going tail first. All turntables for engines should be made on the centre balance plan, so that they can be readily turned on the central pivot, without the use of any gear. In all cases the diameter of the turntables should never be less than from 3 to 5 feet in excess of the longest wheel base to be dealt with, and where this latter is likely to be increased a still greater margin should be taken. This

FIG. 24.

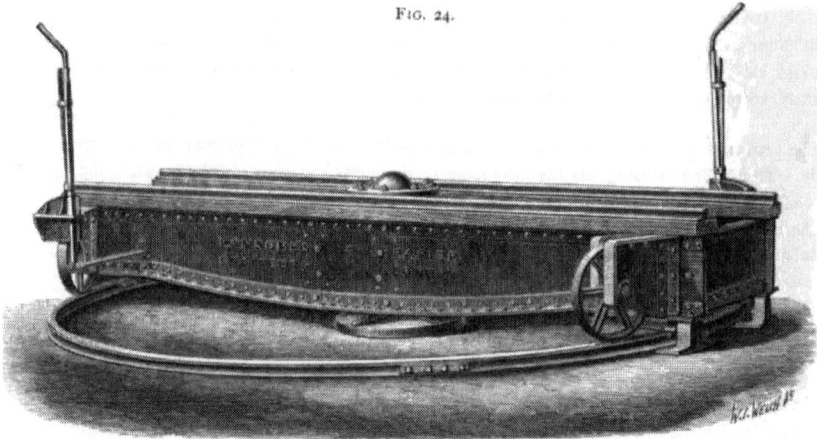

TURNTABLE FOR TANK ENGINES.

margin enables the driver to obtain a better balance of his engine, and therefore to turn it with greater facility. The turntables should consist of two main girders under the rails, well secured to a strong centre piece, having an adjustable steel centre. When the turntable is well made, and the engine is well balanced upon it, one man can turn it by merely pushing it round.

FIG. 25.

TURNTABLE FOR TENDER ENGINES.

For tank engines the turntables may be from 10 to 18 feet diameter (Fig. 24), and for small tender engines from 25 to 40 feet. (Fig. 25.) Up to 18 feet diameter, it is best to have the table made with a wrought-iron pit, as this can be very cheaply fixed. For larger tables a masonry, brick, concrete, or timber pit can be made on the spot.

Sometimes small turntables will be required for carriages and wagons; the use of these should, however, be limited to such confined situations as piers, docks, carriage or goods sheds. (Figs. 26, 27.) In Great Britain turntables are now never laid in the travelling lines.

FIG. 26.

TURNTABLE FOR CARRIAGE SHEDS.

FIG. 27.

TURNTABLE FOR WAGONS.

TRAVERSERS.

When there is not room for points and crossings, the traverser is now almost universally used in preference to turntables, as the means of obtaining access from one pair of metals to another.

FIG. 28.

TRAVERSER FOR CARRIAGES AND WAGONS.

It consists of a low framework of wrought iron mounted on small wheels, arranged to travel on rails laid transversely to the metals of the lines. The sides of the framing are furnished at the bottom with ledges projecting outwards, so as to receive the wheels of the vehicle intended to be transferred from one line to another. (Fig. 28.) The vehicle is wheeled on to the traversing

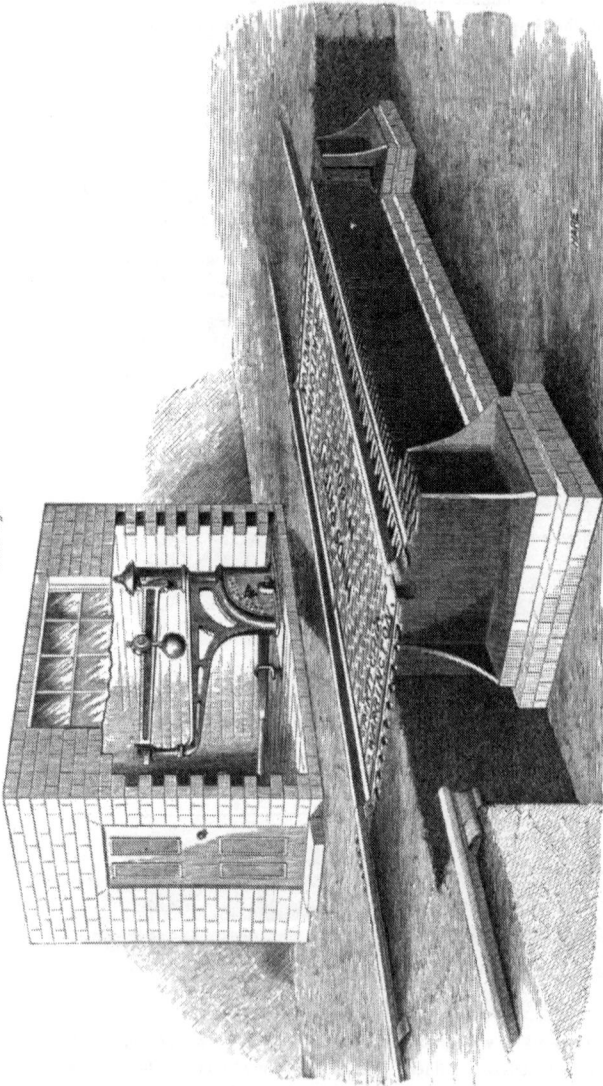

FIG. 29.

WEIGHBRIDGES FOR GOODS STATIONS.

table, which is then pushed in a transverse direction, across so many lines of rail as there may be, until the desired line is reached ; the traverser is then adjusted and held by stops, and the vehicle is wheeled off the traverser on to the adjacent rails.

The important advantage which the traverser has over the turntable consists in the fact that when it is not in use it can be pushed out of the way of the travelling roads, whereas turn-tables have to endure all the hammering of the trains. Also, when turntables are used for transferring carriages from one line to another, there must be at least two tables—in fact, as many as there are roads to be communicated with, whereas one traverser will suffice for any number of roads.

WEIGHBRIDGES.

The convenience of being able to weigh entire vehicles at will is obvious. Weighbridges are made of all sizes, and a good guide as to the size to be adopted will be found to be one and a half times the weight of a fully-loaded goods wagon ; the machinery will thus be always strong enough for its work, and less likely to be deranged in its action. For use abroad weighbridges should always be self-contained, as shown in Fig. 29, and independent of any structural adjustments at destination. The woodcut shows one of Mr. Hind's most recent examples. They should also be fitted with relieving apparatus to support the table on solid bearings when not actually in use, and the scale should be marked in the weights of the country on one side, and in English weights on the other side. Weighbridges should be long enough to receive the wheel base of the longest vehicles on the line, with something to spare. Weighbridges are often made in several sections for weighing locomotives, so as to ascertain exactly the weight on each pair of wheels. In America an extension of the same principle, with some modifications, has been often applied for the purpose of weighing a whole train at once. These are useful luxuries in their places, but quite beyond the practice of cheap railways.

WATER SUPPLY.

For small railways frequent supply of water for the engines is very desirable. If the watering stations are 10 miles apart as an average, the engines should carry water sufficient for 25 or 30 miles, so as to cover contingencies. Watering stations are generally made at those stations which have most traffic, and which involve the longest stoppages, but it often happens that the stations which are the most convenient have the worst water. The quality of water supplied to the boilers is of so much consequence, that it is best to make the engine tanks of ample size, so as to carry considerably more than the minimum supply necessary, and this enables a driver to choose the best water. In some parts of the Cape Colony good water is only to be had at very long intervals of distance ; and to meet this, engines have been constructed as tank locomotives with tenders in addition, so as to enable engines of moderate size to carry water for 60 miles run. It is a common practice to fix the tank at one end of a station for watering the trains going in one direction, and a platform water-crane at the other end of the station fed by pipes from the same tank.

A plain standard water crane is shown at Fig. 30 of 5 inches internal diameter, and suited for 4-inch mains.

When the number of trains is likely to be few, and only a small supply required, a very convenient arrangement is shown in Fig. 31. The tank is mounted on cast-iron columns fixed on masonry or concrete footings, and is fitted with valve leather hose for feeding the engine, and also with a pump for filling the tank. One man can give a supply of 300 gallons per hour, and in case of need two men working in turns could do twice that quantity.

FIG. 30.

FIG. 31.

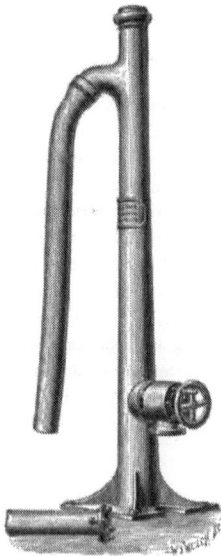

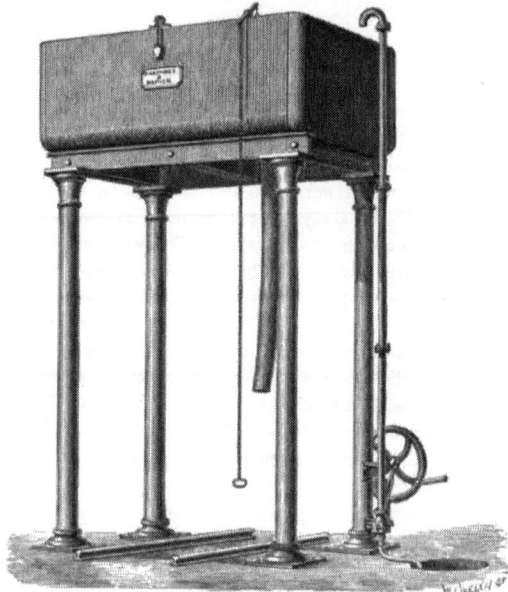

WATER COLUMN FOR PLATFORMS.   WATER TANK, IN CAST IRON.

By an arrangement of a three-way cock, closing the delivery to the tank, the pumps can be made to serve the purpose of a pressure pump for washing out the locomotive boilers, a hose and branch pipe being attached to the nozzle provided for that purpose: these fittings may also be made available in case of fire.

For terminal stations larger tanks will be required, and especially for the depôt station. In these cases it will sometimes be found best to fix the tank on the top of a masonry building prepared to receive it, the building itself being available for other purposes—very often used as a pumping house, and sometimes as a lamp room, or even as a small workshop. In all cases the tanks should be made of cast-iron plates, all planed to one uniform size, so as to be thoroughly interchangeable. When so planed the joints can easily be made with a little thick red-lead paint. The joints may also be made with recesses to receive cement in case of need.

Cast iron is by far the most durable material for water tanks, and is therefore to be preferred. Sometimes, if there is a very long land carriage, the balance of advantage may lie in favour of wrought iron. When made of this material they should always be circular in form, with a

FIG. 32.

FIG. 33.

WATER TANK, IN WROUGHT IRON.

SIGNAL POST FOR STATIONS.

balloon bottom, the tank being supported on an angle iron round its external periphery. The first cost is about the same as for cast-iron tanks of equal capacity. A very convenient form of small wrought-iron tank is shown in Fig. 32. This is very suitable for intermediate stations.

SIGNALS.

For very small railways and low speeds, signals may be dispensed with, if there is very little traffic. The convenience, however, of being able to protect a station yard, at so little expense, far outweighs the trifling economy effected by dispensing with signals. Even though there may be only one engine in steam at a time, wagons moved by hand about the station yard may be in the way, or obstructions may arise from a variety of causes. It is obvious that it is most desirable to have a clear understanding between the station officials and the driver of an approaching train; and this can be so simply secured by signals, that the trifling expense is well laid out. In the crowded traffic of Great Britain, it has been found necessary, not only to secure a clear understanding between stations and drivers, but to

G

ascertain by mechanical contrivances that the various points are all in proper position, before any signals can be given. This is called the interlocking system, and it has been carried out at very great cost and to complete perfection. The regulations of the Board of Trade on the subject are very stringent, though perhaps not unnecessarily so. Some relaxation is allowed in the case of light railways.

For light railways abroad interlocking of signals and points is as yet quite unnecessary. A signal mast of semaphore type (Fig. 33) with two arms, one for each direction, may suffice if the approaches are such that it can be seen for a distance of half a mile or so. If curves or gradients prevent the signal being seen far enough off, it would be better to use distant signals placed 500 yards from the station in each direction, and worked by a lever at the station. In most cases the station signal might be dispensed with, if distant signals are carefully and regularly used.

On this point it is necessary further to add a caution, that if signals are fixed they ought to be constantly used and attended to. The author has known instances where the signals were not used, but allowed to stand always at danger or at safety. If habitually regarded as of no importance, they will certainly be unobserved or neglected at the moment of danger.

TELEGRAPH.

The telegraph is such a useful adjunct, that it may be said to be almost a necessary of life to a railway. In England the block system is now generally adopted, and under this method of working, all the movements of trains are controlled and directed by telegraph. On lines of railway abroad, the telegraph is desirable for the opposite reason. The fewness of the trains requires the telegraph as a means of communication, owing to the long intervals between trains. Thus in all circumstances the telegraph plays an important part.

The simplest kind is the magnetic alphabetical. The cheapest kind is the single needle electric; and the best is the electric ink recording telegraph.

The following estimates of telegraph materials have been kindly furnished by Messrs. Siemens Brothers, of Westminster. It is therefore needless to add that the materials are of the best known kinds and of best possible manufacture.

If wooden poles are intended to be used they will probably be furnished in the locality. Their use must depend on many circumstances, such as the suitability of the wood obtainable, its first cost, probable durability, &c., &c.; suffice it to say, that unless the saving of first outlay is imperative, it will be found better to use iron poles, the inconvenience of replacing wooden poles being very great and very costly when interruptions of the use of the telegraph are caused thereby.

In sending poles from this country, it should be borne in mind that wooden poles cost a great deal more than iron for freight and transport. Messrs. Siemens' patent iron poles are usually made about 20 feet high, and consist of a dished or buckled foot plate of wrought iron, to which the lower portion of the pole is attached by four bolts. The lower portion is made

FIG. 34.

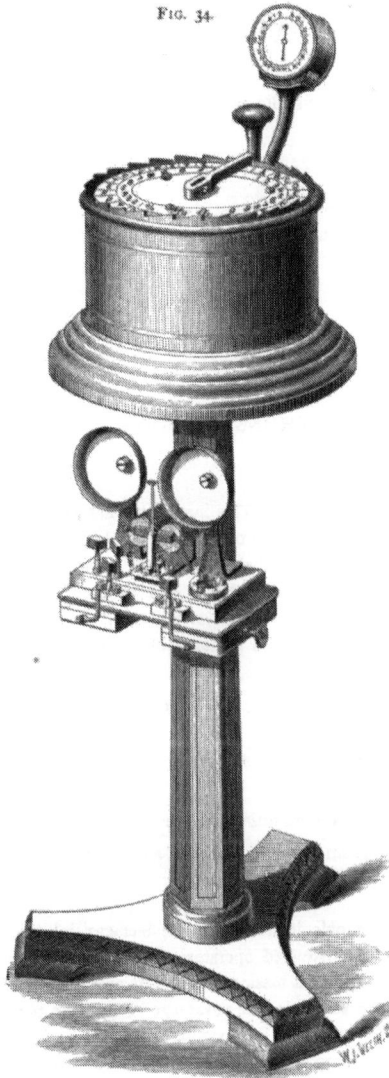

TELEGRAPH—MAGNETIC DIAL.

of cast iron for a height of from 2 to 4 feet above ground. Cast iron is not subject to corrosion at the ground-line as wrought iron and timber are, and it therefore forms the best

G 2

material for the lower part of the pole. The upper part is made of wrought iron, and is firmly secured in the lower portion by cement. The top of the pole is finished with a rod 2 feet high, forming a lightning conductor.

The number of posts per mile is usually eighteen ordinary or intermediate, and three stretching or corner posts.

Cost for an ordinary line :—

18 ordinary and 3 stretching posts cost about £30
Insulators and wire, per mile 7
Spare materials 1
Total per mile £38

Tools for fixing, suitable for any length of line up to 50 miles, price £43.

TELEGRAPH INSTRUMENTS.

Instruments are of three principal kinds :—

I. Magnetic alphabetical telegraph. Fig. 34. This instrument has two dials with the letters of the alphabet written round them. The one dial is furnished with a handle, and is the means of transmitting signals, and the other has a pointer by means of which signals are received. This instrument is so simple that it can be used by anyone after a few minutes' practice.* It is quite independent of batteries, and consequent attention; it is also very strong in all its parts, and is not liable to be deranged, even by rough usage.

Cost per station, complete with alarums, &c., &c., £33.

II. Single-needle electric instruments, with batteries and all apparatus.

Cost for terminal stations, £29.

Intermediate stations £5 more for extra fittings required.

This kind of telegraph requires attention to batteries, and facility in its manipulation can only be acquired by considerable practice.

III. By far the best form of telegraphic instrument is the Siemens' ink-recording telegraph. This instrument also requires instructed operators, and attention to batteries. As its name implies, it keeps a record of all messages sent. Its cost is £52 for an end station, and £57 for an intermediate station. The advantage of having a permanent record of all messages passing through an office is obviously very great, as it affords a constant means of reference for the purpose of tracing mistakes or defects in the working,

---

* The telephone will probably soon come into use as a substitute for the dial telegraph. It enables one person actually to speak to another at a distance of hundreds of miles, and its movements are so perfect that even the voice of the speaker can be recognized. The instrument is still in its infancy.

the record being indisputable evidence of the message as received. For the general purposes of a light railway, however, the magnetic telegraph will probably be found to be the most convenient.

Cost of materials required for 20 miles of single wire magnetic telegraph (exclusive of posts) with two terminal and three intermediate stations :—

20 miles of insulators and wire, at £7	£140
Tools for fixing	17
Magnetic dial instruments for two terminal stations, at £32	64
Magnetic dial instruments for three intermediate stations, at £33 ..	99
Spare materials	20
Total for 20 miles	£340
Total per mile, exclusive of posts, but inclusive of instruments ..	£17

Cost of iron posts, £25 to £30 per mile.

Telegraph, with needle instruments and batteries, would cost for such a line about the same as the above ; and telegraph with recording instruments about £22 per mile, exclusive of poles. In every case several copies of printed instructions are supplied with the apparatus, giving the fullest details of working and management.

STATION BUILDINGS AND STATION FITTINGS.

On many lines in America, the expense of intermediate stations is entirely saved by adopting the practice of stopping the trains at certain roads, and giving out tickets by the guard in the train itself. In such cases, passengers join the trains precisely in the same way as was common in the old coaching days, when the custom was to await the passing of the coach at the cross-roads. This, of course, is the simplest way of getting rid of the expense of stations. A first step would naturally be to provide a shed as shelter from the weather, and when the traffic might increase, and a resident officer of the railway might become desirable, more accommodation would have to be provided. Precisely how little will suffice, or how much can be afforded, must depend on the circumstances at the time. In England we are accustomed to stations of considerable luxury, with refreshment and waiting rooms of many varieties for different classes of passengers ; and at many stations these are found on both sides of the line. This, however, is the outcome of years of successful traffic. When railways were first made in England, the stations were of a very inexpensive character ; in most cases the station consisted of a wooden hut for giving out tickets, and a little dry gravel for a platform. More than this is not absolutely required ; if the rolling stock be made with floors not more than 2 feet from rail level, access to the carriages can be had quite readily from a low platform formed of gravel about six inches higher than the ballast. A platform of some kind is useful as indicating distinctly where passengers are intended to alight from or to enter the trains. Stations should be limited to a very minimum of outlay until the line has paid 5 or 6 per cent. dividend. In England the Board of Trade now insists on all stations having separate waiting rooms for different classes and sexes; and high platforms long enough for the longest trains, even though there may be only one or two passenger vehicles in the train. Formerly these things were not considered at all necessary. Certainly in making railways in new districts such extensive accommodation should not be thought of in the first instance.

FIG. 35.

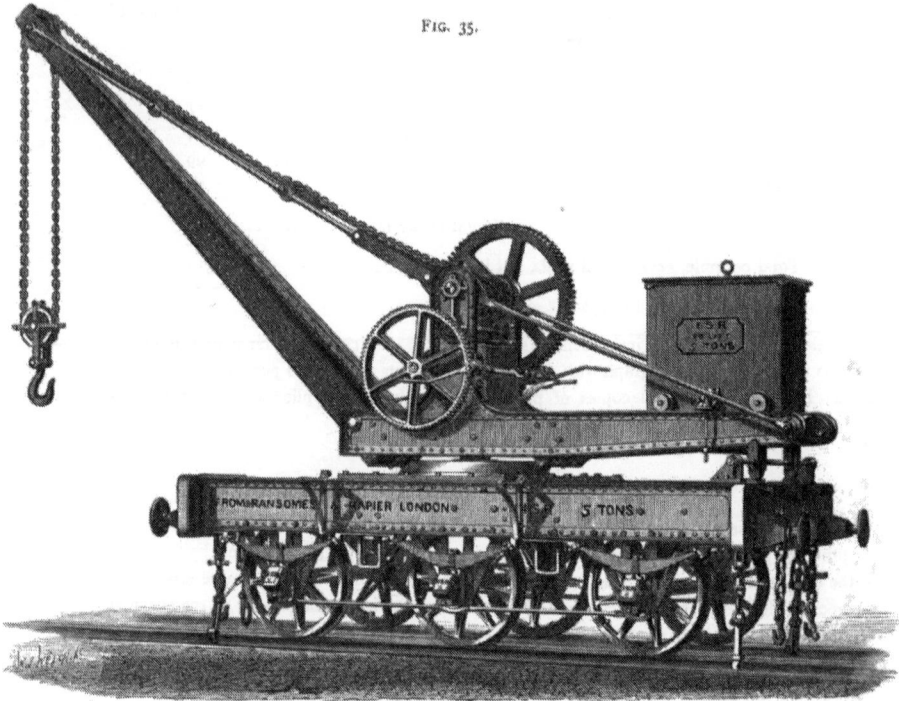

PORTABLE CRANE.

In the estimates a small sum is set down for station fittings to cover the cost of ticket cases and drawers, dating stamps, stationery, clocks, station furniture, a hand-pump, luggage barrows, platform weighing-machines, and other necessaries. For loading goods, a loading bank made up to the floor level of the trucks will be economical, and will save much time in loading. If goods in heavy bales are frequent, a small wharf-crane will also be of great service, the size being determined by the goods offering or likely to offer. In this country, roadside stations have been fitted with 10-ton cranes when there was no object of that weight in the neighbourhood except the church steeple.

Cranes on wheels are very useful, because they can be borrowed by one station from another, and they are also serviceable in case of accident or breakdown.

A very convenient form of portable or breakdown crane is shown in Fig. 35. The side frames are made of wrought iron, and the tail for carrying the balance box is formed in one piece with the sides. The jib, crane post, and carriage frame are also of wrought iron. The wheels and axles, axle boxes, springs, buffers, and couplings should all be of the normal type in use on the line, so that needful repairs can be immediately effected.

Whenever the goods traffic is such, in nature and extent, as to require goods warehouses, they should be fitted with small cranes to lift 30 cwt., placed near enough to each other to command the whole of the floor.

WORKSHOPS.

For a short line of railway near to a good town, a small workshop without machines may suffice; but if there are no other means of getting machine work done, the railway must have its own complete establishment.

FIG. 36.

WORKSHOP ENGINE.

The size of the engine wheels will determine the size of the lathe required for re-turning them. If the line is very short, or the supply of rolling stock is very limited, a large lathe for general purposes may suffice; for example, a 16-inch lathe will re-turn wheels of 32 inches diameter over the flanges, and will also be useful for many other purposes. Next after a lathe, the most serviceable tool is the radial drilling machine; and, however small a workshop plant may be, it should always include one or more of these most useful machines. The best type of engine for driving workshop machinery is the horizontal engine shown in Fig. 36. This engine can be fixed on masonry; or where this is expensive, two long and heavy blocks of timber will suffice. The power is communicated to the main shafting by a strap; the engine should be placed at a distance from the main shafting equal to three or four times the diameter of the fly-wheel—the longer the strap, the greater its durability. The most convenient type of boiler is the multitubular. It is easily fixed, and all its parts are easily accessible. (Fig. 37.) Lists of machine tools with steam engines and boilers for driving them, and also smiths' fires and fans, &c., &c., are subjoined, particulars being given of four different sizes. Vide Appendix II.

FIG. 37.

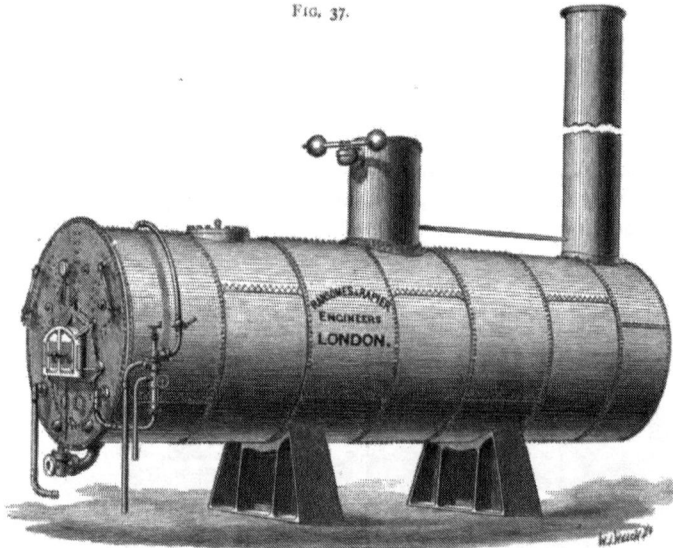

WORKSHOP BOILER.

The workshop itself may be constructed of such materials as may be available on the spot. Or workshops constructed of iron may be sent from this country. In districts where skilled labour is scarce an iron workshop will be found to be economical. It can be completely erected in this country, and thoroughly marked for re-erection at its destination. It can be all put together by unskilled labourers if the marking has been well attended to. In addition to marking by letters and figures, frequent marks consisting only of different numbers of stripes across important joints of the work will be found to be a very great help. Different widths of stripes can be used, and also different colours.

ROLLING STOCK—LOCOMOTIVES.

It will have been observed that in the preceding pages the tank engine has been almost exclusively advocated.

For cheap railways the tank engine is naturally the favourite type; because, in places where speed is not required, maximum adhesion is generally the chief object, and a tank engine which carries its supply of water and coals on the same wheels as the engine itself, has an advantage in adhesion and hauling power. The old practice of putting a sack on a horse's back to assist him in drawing a load up hill is strikingly exemplified in the steam horse.

As traffic increases, improved speed, and longer runs without stopping, will probably be required. This demand may be best complied with, by running passenger trains distinct from goods, and thus the higher speed will be commonly wanted for lighter loads. As this phase. develops, it will be found advantageous to introduce at a further stage tender engines. For example, given a line with tank engines six wheels coupled, and weighing 12 tons, running 20 miles an hour, and working a mixed traffic at a journey speed of 15 miles an hour; an easy improvement may be made by adopting an engine, with four coupled and two trailing wheels, weighing 9 tons, and a tender weighing 5 tons, and a running speed of 25 miles an hour; and by making fewer stoppages the journey speed might be very nearly equal to that, say 23 miles an hour. An improvement of 50 per cent. in journey speed, for light trains, may thus be effected without distressing the road, if the weight of the engine and its number of coupled wheels be reduced as suggested above. Economy in the cost of engines should be sought only in the reduction of size. Small engines should have copper fire-boxes, Low Moor boilers, and steel tyres, and should be in all respects of as good materials and workmanship as larger ones; the best engines will prove to be the cheapest. In the estimates full particulars are given as to dimensions, capability, and cost of various sizes, by different makers; in fact, a succession of short specifications.

CARRIAGES AND WAGONS.

The most economical kind of carriage, both in cost and in traffic working, is the omnibus type, with seats at the sides and arm-rests at occasional intervals to prevent undue monopoly of space : the doors being at the ends and the floors continued so as to form end platforms; access can be had at all times to all parts of the train. This plan has so many advantages in working that it is becoming very generally adopted. In some cases, however, carriages divided into compartments may be required, and the cost of such may be taken as 10 per cent. more than the other kind. So large a proportion of the railways now under construction are in hot climates, that the estimated prices are intended to cover the cost of double roofs and also of louvre windows in addition to glass windows, the seats being both fitted with cane for hot weather and furnished with cushions for winter use.

For branch line traffic it is not necessary to have such commodious and expensive vehicles as for longer distances. A short journey even in a cramped position may be borne, whilst for long journeys the maximum of comfort is desirable.

Fig. 38 gives a view of the interior of first-class carriages for railways of 2 ft. 6 in. to 3 ft. 6 in. gauge; the interior dimensions being 18 feet long by 5 ft. 6 in. to 6 ft. 6 in. wide, there is ample space for comfortable seats and for a good gangway along the centre.

Fig. 39 gives the exterior view of the same carriage. It weighs $2\frac{3}{4}$ tons empty, and with 20 passengers about 4 tons. Similar carriages are made weighing 2 tons empty, and 3 tons with 12 or 14 passengers.

H

FIG. 38.

INTERIOR OF FIRST-CLASS CARRIAGE.

FIG. 39.

EXTERIOR OF FIRST-CLASS CARRIAGE.

Fig. 40 represents a second or third class carriage. The interior is arranged in a similar manner to the one last described. On light lines, guards' lockers may with advantage be fixed, the second-class carriages, for the reception of parcels or mails.

The above examples are from the portfolio of the Bristol Carriage and Wagon Company.

The second and third class carriages may be very conveniently fitted with brakes fixed on the end platforms, as shown in the figure, and by making up the train with two brakes together, one guard can easily manipulate both. Particulars are given for three classes of carriage, as local

Fig. 40.

EXTERIOR OF SECOND-CLASS CARRIAGE.

circumstances can alone decide whether two classes are sufficient or not. It is well known that there are five classes in India; and in China three classes have been found to be desirable, by reason of the great disparity between the fares which first-class passengers are willing to pay, and those which third-class passengers can be induced to pay. In some places one class may suffice, and if so, so much the better.

Central buffers and self-acting couplings afford such facilities that they are to be preferred for low speeds; but, for very high speeds, side buffers give greater steadiness, as the vehicles can then be tightly screwed up buffer to buffer, and each carriage is steadied by its neighbours. Means have been contrived for tightening up central buffers, but even when so adjusted the carriages are not protected against lateral oscillation as they are by side buffers.

The wheels both of carriages and wagons are sometimes as small as 18 or 20 inches in diameter. Steel tyres are to be preferred, but chilled wheels may be used if extreme economy in first cost be imperative. Solid steel disc wheels have also been used with success. They are somewhat cheaper than wheels with steel tyres; and when worn, the flanges can be turned off and rolled tyres can then be put on.

If the wheels are larger than 2 feet they should be made to come up through the floor, but the necessary hoods to cover them are rather inconvenient in narrow-gauge stock if the seats are arranged longitudinally. Nevertheless it is most desirable to keep the floors as low as possible, so as to secure a low centre of gravity, and consequently a good angle of stability. In this way light narrow-gauge stock can be made to run just as steadily and safely as that on wider railways. The width of the inside of the carriages may be equal to twice the gauge: this rule has been slightly exceeded without any inconvenient results.

The underframes both of carriages and wagons may be made a little more than half the dimensions for full-sized stock. Carriages to accommodate twenty passengers need not weigh more than 2¾ tons empty.

Generally speaking, wagons for light railways have been made unnecessarily strong and heavy. The wagons shown in Fig. 6 measure 10 feet by 5 feet by 2 feet, and weigh only 24 cwt. empty. They are, nevertheless, strong enough for a load of 4 tons.

The wagon indicated in Fig. 41 weighs only 16 cwt., but is equal to a load of $2\frac{1}{4}$ tons. It has low sides, and a floor space of 7 feet 6 inches by 4 feet, so as to admit of carrying two hogsheads of sugar on a narrow-gauge line. The diameter of the wheels is 16 inches, and the floor is only 22 inches from rail level. The sides are hinged so as to fall down, and this, coupled with the lowness of the floor, renders the wagons very easy to load and unload, and also makes them very well adapted for ballasting. The ends are supported by stanchions, fixed in sockets, so that they can be taken out. Two or three wagons can thus be converted into one for carrying rails or timber.

Whenever small wagons will carry the traffic, they are much to be preferred to larger ones, because a small wagon properly constructed will carry a greater proportion of paying load to its dead weight than a larger one; also there is less waste in the loading of goods of different traders in separate wagons, and the average loads actually sent approximate more nearly to the full loads. Moreover, a large number of small wagons enables a greater number of persons to have some supply of stock, than a smaller number of larger wagons.

FIG. 41.

SMALL WAGONS.

In the succeeding estimates prices are given for low-sided, high-sided, and covered goods wagons. Platform wagons with stanchions may be estimated at the same cost as the low-sided wagons. Prices are also given for timber wagons, cattle trucks, and horse boxes. The relative numbers of the vehicles of the different kinds must vary according to the local requirements. In all these estimates only sufficient stock has been suggested for the first opening of the line, with a view to enable remunerative business to be commenced with the least practicable outlay, the eventual full supply of rolling stock being left to be determined by the light of experience.

Before leaving the subject of rolling stock, reference should be made to the necessity of providing some means for enabling vehicles of long wheel-base to pass very sharp curves with ease and safety. Of all the contrivances for this purpose Mr. Cleminson's is so decidedly the simplest and best that an account of it is subjoined. See Appendix III. The author would only remark that as long as small vehicles of short wheel-base will do the work, they are to be preferred; but if very sharp curves are necessary, then Mr. Cleminson's plan may be adopted, with great advantage, for vehicles of such length as may necessitate a long wheel-base.

CHAPTER VII.

In some instances light railways have not been successful, and much disappointment has ensued. It is right to point out that such disappointment may arise, and indeed may be expected. For instance, if the route of a long line of railway is decided by military considerations, rather than with the object of accommodating the ordinary traffic from town to town, financial success can scarcely be looked for. It may be that the line does not go near the towns, or that there are no roads by which the trade of the country can gain access to the railway, or it may even be that the railway is made where there is no trade at all.

Again, in some cases it has been said that the light railway was unable to take sufficient train loads to be economically worked, or that it was unable to accommodate all the traffic offering.

The latter proposition may be admitted if the desire is to crowd all the traffic of the day into one or two trains; and the former allegation may also be true, if the railway is made only of narrow gauge, and with light rails, but with all the rolling stock nevertheless following closely upon English standards of dimension and weight, and if the method of working be such as to involve an approach to English speed of travelling, and to English cost of maintenance of road and general service.

But if a railway be made between existing centres of population, or business, and the route be well selected, and the railway be made on a scale suitable for the traffic immediately expected, and if care be taken that the cost of the service shall not exceed the remuneration immediately obtainable for it—in short, if a railway be made and conducted on the same principles of mutual advantage as any other business, success is certain.

In conclusion, the author would only add—

1. Let the size of railway best adapted for the purpose in view be well considered and determined.

2. Let the original intention be religiously adhered to; the beginning of success may be easily marred by allowing the railway to be overtaxed with speed or heavy weights.

3. When a profit appears, care should be taken to let well alone.

4. Increase of speed, enlargement of stations, and extension of works and accommodation should all be postponed until the line has not only paid a dividend on its first outlay, but until it has accumulated a reserve fund for improvements, or has at least shown that prospective revenue is well in sight to justify increased expenditure.

I

Estimates for materials and fittings, probably required from England, are given in the following pages. The different examples are given, beginning from the smallest Pioneer size with an engine weighing 2 tons, and rails of 12 lbs. per yard, up to engines of 18 tons, and rails of 45 lbs. per yard.

These estimates have been prepared with a due regard to the maximum of efficiency in all details, and all prices have a considerable margin beyond the very low figures current at present (January, 1878); all figures are put down so that anyone may reasonably hope to obtain the various requirements at prices "within the estimate," and so avoid the disappointment which might result from too close a statement.

Similarly the duty of the engines, both as regards speed and power, is given at what they can be well depended upon to do constantly, from one day to another, and under average conditions of climate. Test duty is not economical every-day duty.

As in every case there will be some deviation in detail from the particulars set down, the cost of every item has been given with as much detail as possible. Thus, from any of the estimates, an estimate for a line of different length or different number of stations can be readily calculated; or, where funds permit, some may prefer to use the size of engine set down in one estimate, and the permanent way of the next heavier estimate.

To facilitate the use of the figures here given, a few pages of schedules are appended with all figures left blank, so that estimates of exact requirements can be readily compiled, omitting or varying any items according to circumstances. These pages are so arranged that the book may be opened at the estimate intended to be taken as the basis, and the pages to be written on can at the same time be opened out, so that both can be in full view at once. In making such estimates, the figures should be entered in soft pencil until finally decided, and then they may be entered in ink. The figures may afterwards be copied on to the inner portion of the sheet, and preserved in the book as a record; and the outer portion can be cut off for transmission to correspondents.

If a credit be opened with a well-known banker or merchant, and a consulting engineer of good position be requested to inspect purchases, there will be no difficulty in obtaining the appliances required to secure any desired result. Only let care be taken to consider well the duty to be done. "It is not necessary to provide an elephant to draw a cart."

TABLE II.—TABLE GIVING COMPARATIVE SUMMARY OF LEADING PARTICULARS
FOR LIGHT RAILWAYS OF EIGHT DIFFERENT SIZES

(For which details are given in the following Pages).

No.		SIZES.							
	—	No. 1.	No. 2.	No. 3.	No. 4.	No. 5.	No. 6.	No. 7.	No. 8.
1	Weight of Rails per Yard (Vignoles type)	12 lbs.	15 lbs.	15 lbs.	20 lbs.	25 lbs.	30 lbs.	40 lbs.	45 lbs.
2	Weight of Engine	2 tons	4 tons	4 tons	7 tons	9 tons	12 tons	15 tons	18 tons
3	Number of Wheels	4	4	6	6	6	6	6	6
4	Number of Wheels Coupled	4	4	4	4	6	6	4	6
5	Greatest Weight on any Wheel ..	10 cwt.	20 cwt.	17 cwt.	25 cwt.	30 cwt.	2 tons	3 tons	3½ tons
6	Total Wheel Base of Engine ..	3 ft.	4 ft.	5 ft. 6 in.	7 ft. 5 in.	8 ft. 6 in.	9 ft. 6 in.	11 ft.2 in.	10 ft. 10 in.
7	Length of Fixed Wheel Base, with 4 Wheels Coupled	3 ft.	4 ft.	3 ft.*	3 ft.6 in.†	4 ft. 6 in.	5 ft. 0 in.	6 ft.6 in.*	6 ft. 6 in.
8	Diameter of Driving Wheels	1 ft. 6 in.	1 ft. 8 in.	1 ft.10 in.	2 ft.	2 ft. 3 in.	2 ft. 6 in.	3 ft. 6 in.	3 ft.
9	Speed with Light Loads .. per hour	12 miles	12 miles	12 miles	14 miles	16 miles	18 miles	23 miles	20 miles
10	Speed on a Level .. .. .. ,,	miles. tons. 7 34	miles. tons. 8 30	miles. tons. 8 60	miles. tons. 8 80	miles. tons. 9 120	miles. tons. 10 160	miles. tons. 11 210	miles. tons. 10 to 12 250
11	Speed on Gradient, 1 in 100 .. ,,	5 12	5 16	5 30	5 36	6 55	7 75	8 80	7 to 8 100
12	Speed on Gradient, 1 in 35 .. ,,	3 7	3 11	3 18	3 22	4 33	4 45	5¾ 45	4 to 5 50
13	Cylinder .. (diameter and stroke)	in. in. 5 × 6	in. in. 5½ × 7	in. in. 6 × 8	in. in. 6 × 12	in. in. 8 × 12	in. in. 9 × 14	in. in. 11½ 17	in. in. 10½ × 18
14	Cost of Materials, Plant and Rolling Stock per Mile, for a Line 20 Miles long..	£500	£550	£750	£1,279	£1,735	£1,892	£2,400	£2,700

* With Sliding Axle Box.   † With Patent Axle Box.   ‡ When made as a Tank Engine.

SIZE No. 1.—(*See Estimate No. 1.*)

Estimate of Materials and Fittings

FOR

FIVE MILES OF RAILWAY FOR STEAM CARRIAGE TRAFFIC,
WITHOUT ANY STATIONS.

DUTY REQUIRED TO BE DONE.

1. Maximum train load required to be taken at 7 miles an hour on a level = 24 tons.
2. „   „   „   „   5 „   „   up 1 in 100 = 12 „
3. „   „   „   „   3 „   „   „ 1 in 35 = 7 „
4. Speed with light load = 14 miles an hour.

GAUGE, 2 ft. o in. to 3 ft. o in.

FIG. 2.

DESCRIPTION OF ENGINE SUITABLE FOR THE REQUIREMENTS.

1. Weight in working order, 2 tons, on 4 wheels, 1 ft. 6 in. diameter, all coupled.
2. Cylinders, 5 in. × 6 in. Heating surface, 60 square feet. Grate area, 2 square feet.
3. Greatest weight on any wheel 10 cwt. Rails, 12 lbs. per yard. Gauge, 2 ft. to 3 ft. o in.
4. Fixed wheel-base of Engine, 3 ft.

ESTIMATE No. 1—(*for Size No.* 1).

Measurement.	Dead Weight.			£ s. d.	£ s. d.
..	100	Rails (12 lbs. per yard), with fish plates	Tons, 100 at	9 0 0	.. 900 0 0
..	7	Fish bolts, dog spikes, and fang bolts	,, 7 ,,	24 0 0	.. 168 0 0
146	..	Sleepers, 7 in. × 2½ in. × 4 ft. (2400 per mile) .. .. ..No., 12,000 ,,		0 1 2	.. 700 0 0
..	2	Points, 6 ft. tongues ; 8 ft. 6 in. stock rails	Sets, 6 ,,	7 0 0	.. 42 0 0
..	1	Crossings	,, 6 ,,	4 0 0	.. 24 0 0
..	2	Platelayers' tools and trollies		50 0 0	.. 50 0 0
1	..	Hand-pump, with brass barrels, pipes, &c.	No., 1 ,,	30 0 0	.. 30 0 0
..	3	Tanks, of cast-iron plates, with planed joints, to hold 500 gallons, with pipes, valves, leather hose, &c.	,, 1 ,,	40 0 0	.. 40 0 0
..	1	Sundry tools		30 0 0	.. 30 0 0
2	..	Sundry stores, oil and tallow boxes, rope, chain, spunyarn, padlocks, screw jacks, &c., &c.		100 0 0	.. 100 0 0
4	1	Engines, copper fire-box, brass tubes, steel tyres, with lamps and tools complete'	No., 1 ,,	350 0 0	.. 350 0 0
1	1	Tender for coals and water, with seats for 6 persons	,, 1 ,,	80 0 0	.. 80 0 0
1	..	Engine duplicates		50 0 0	.. 50 0 0
4	..	First-class carriages, to hold 8 passengers	No., 1 ,,	150 0 0	.. 150 0 0
17	..	Open goods wagons, 5 ft. × 3 ft. 6 in. × 1 ft. 3 in. ; 7 cwt. empty ; 30 cwt. full (gauge, 2 ft. 6 in.)	,, 25 ,,	18 0 0	.. 450 0 0
1	..	Extra sets of wheels and axles	,, 4 ,,	8 0 0	.. 32 0 0
..	2	Sundry dogs and bolts, for timber bridges, &c.	Tons, 2 ,,	20 0 0	.. 40 0 0
	120	Tons dead weight.			
177		Tons measurement.			

TOTAL COST of all materials and fittings for 5 miles (including sleepers) £3216 0 0

,, ,, ,, per mile £643 4 0

For a line 10 miles long, with 2 engines, the cost of materials and fittings would be, per mile £550 0 0

,, 20 ,, ,, 3 ,, ,, ,, ,, £500 0 0

ITEMS WHICH MAY OR MAY NOT BE REQUIRED.

									£ s. d.
Cost of plain girders for 40 ft. span, with rails on top, and cross stretchers						.. .. per pair	90 0 0
,, ,, 30 ,, ,, ,, ,, ,, ... ,,									50 0 0
,, ,, 20 ,, ,, ,, ,, ,, ... ,,									30 0 0
,, rolled joists 15 ,, ,, ,, ,, ,, ... ,,									18 0 0

SIZE No. 2.—(*See Estimate No. 2.*)

Estimate of Materials and Fittings

FOR

TEN MILES OF RAILWAY FOR STEAM CARRIAGE TRAFFIC,
WITHOUT ANY STATIONS.

DUTY REQUIRED TO BE DONE.

1. Maximum train load required to be taken at 8 miles an hour on a level = 30 tons.
2. ,,   ,,   ,,   ,,   5 ,,   ,,   up 1 in 100 = 16 ,,
3. ,,   ,,   ,,   ,,   3 ,,   ,,   ,, 1 in 35 = 9 ,,
4. Speed with light loads = 16 miles an hour.

No—TE. This Steam Carriage is also made with all wheels coupled, its tractive force being thereby nearly doubled.

GAUGE, 2 ft. 0 in. to 3 ft. 6 in.

FIG. 3.

DESCRIPTION OF ENGINE SUITABLE FOR THE REQUIREMENTS.

1. Weight in working order, 4 tons, on 4 wheels, 1 ft. 8 in. diameter.
2. Cylinders, 5½ in. × 7 in. Heating surface, 85 square feet. Grate area, 2½ square feet.
3. Greatest weight on any wheel, 20 cwt. Rails, 15 lbs. per yard. Gauge, 2 ft. to 3 ft. 6 in.
4. Fixed wheel-base of Engine, 4 ft.

ESTIMATE No. 2—(for Size No. 2).

Tons for Freight. Measurement.	Tons for Freight. Dead Weight.	Description	Quantity	£ s. d.	£ s. d.
..	250	Rails (15 lbs. per yard), with fish plates	Tons, 250 at	8 10 0	.. 2,125 0 0
..	20	Fish bolts, dog spikes, and fang bolts	,, 20 ,,	24 0 0	.. 480 0 0
364	..	Sleepers, 7 in. × 2½ in. × 5 ft. 6 in. (for 3 ft. gauge), 2,400 per mile	No., 24,000 ,,	0 1 6	.. 1,800 0 0
..	3	Points, 6 ft. tongues ; 8 ft. 6 in. stock rails	Sets, 8 ,,	7 0 0	.. 56 0 0
..	2	Crossings	,, 8 ,,	4 0 0	.. 32 0 0
2	..	Platelayers' tools and trollies			.. 60 0 0
1	..	Hand-pumps, with brass barrel, pipes, &c.	No., 1 ,,	30 0 0	.. 30 0 0
..	3	Tanks, of cast-iron plates, with planed joints, to hold 500 gallons, with pipes, valves, leather hose, &c.		40 0 0	.. 40 0 0
..	1	Sundry tools			.. 30 0 0
2	..	Sundry stores, oil and tallow boxes, rope, chain, spunyarn, padlocks, screw jacks, &c., &c.		100 0 0	.. 100 0 0
6	2	Engines, copper fire-box, brass tubes, steel tyres, with lamps and tools complete	No., 1 ,,	420 0 0	.. 420 0 0
		Extra for making the engine into a close carriage with windows		60 0 0	.. 60 0 0
1	..	Engine duplicates			.. 50 0 0
20	..	Open goods wagons, 7 ft. 6 in. × 4 ft. × 1 ft. 6 in. ; weight, empty, 16 cwt. ; full, 3 tons	,, 10 ,,	30 0 0	.. 300 0 0
		Extra sets of wheels and axles	,, 4 ,,	20 0 0	.. 80 0 0
8	..	Open goods wagons, 5 ft. × 3 ft. 6 in. × 1 ft. 6 in. ; 7 cwt. empty, 30 cwt. full (2 ft. 6 in. gauge)	,, 10 ,,	18 0 0	.. 180 0 0
1	..	Extra sets of wheels and axles	,, 4 ,,	8 0 0	.. 32 0 0
..	3	Sundry dogs and bolts for timber bridges, &c.	Tons, 3 ,,	20 0 0	.. 60 0 0
	284	Tons dead weight.			
405		Tons measurement.			

TOTAL COST of all materials and fittings for 10 miles (including sleepers) £5,935 0 0

					£ s. d.
,,	,,	,,	,,	per mile	£593 10 0
,,	,,	,,	,,	for 20 miles .. .. .. per mile	£550 0 0

ITEMS WHICH MAY OR MAY NOT BE REQUIRED.

					£ s. d.
Cost of plain girders for 40 ft. span, with rails on top and cross stretchers .. .. .. per pair	90 0 0
,, ,, 30 ,, ,, ,, ,, ... ,,	50 0 0
,, ,, 20 ,, ,, ,, ,, ... ,,	30 0 0
,, rolled joists ,, 15 ,, ,, ,, ,, ... ,,	18 0 0

SIZE No. 3.—(*See Estimate No. 3.*)

Estimate of Materials and Fittings

FOR

TEN MILES OF RAILWAY FOR LIGHT MINERAL TRAFFIC,

WITHOUT ANY STATIONS.

DUTY REQUIRED TO BE DONE.

1. Maximum train load required to be taken at 8 miles an hour on a level = 50 tons.
2.   „   „   „   „   5 „   „   up 1 in 100 = 25 „
3.   „   „   „   „   3 „   „   „ 1 in 35 = 15 „
4. Speed with light loads = 16 miles an hour.

GAUGE, 2 ft. 0 in. to 3 ft. 6 in.

FIG. 4.

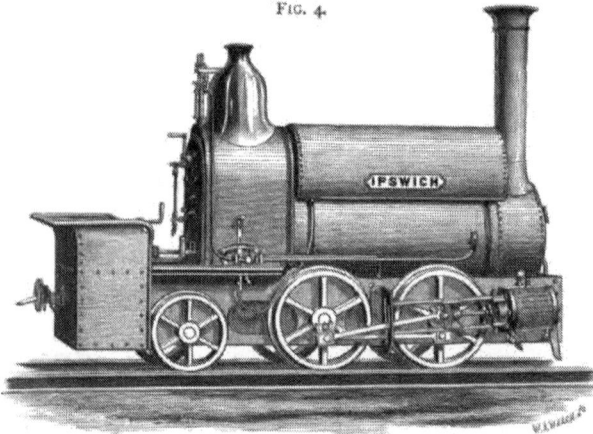

DESCRIPTION OF ENGINE SUITABLE FOR THE REQUIREMENTS.

1. Weight in working order, 4 tons, on 6 wheels, 1 ft. 10 in. diameter, 4 wheels coupled.
2. Cylinders, 6½ in. × 8 in.   Heating surface, 110 square feet.   Grate area, 2½ square feet.
3. Greatest weight on any wheel, 17 cwt.   Rails, 15 lbs. per yard.   Gauge, 2 ft. to 3 ft. 6 in.
4. Fixed wheel-base of Engines when made with six wheels coupled, 5 ft. 6 in.
5. Fixed wheel-base of Engines when made with four wheels coupled, and a "bogie," 3 ft. 6 in.

Tons for Freight.				£ s. d.	£ s. d.
Measurement.	Dead Weight.				

ESTIMATE No. 3—(*for Size No. 3*).

Measurement.	Dead Weight.	Item	Quantity	£ s. d.	£ s. d.
..	250	Rails, 15 lbs. per yard, with fish plates	Tons, 250 at	8 10 0 ..	2,125 0 0
..	20	Fish bolts, dog spikes, and fang bolts	,, 20 ,,	24 0 0 ..	480 0 0
401	..	Sleepers, 7 in. × 2½ in. × 6 ft. 0 in. (2400 per mile)	No., 24,000 ,,	0 1 6 ..	1,800 0 0
..	6	Points, 6 ft. tongues ; 8 ft. 6 in. stock rails	Sets, 15 ,,	8 0 0 ..	120 0 0
..	4	Crossings	,, 15 ,,	5 0 0 ..	75 0 0
2	..	Platelayers' tools and trollies		60 0 0 ..	60 0 0
3	..	Weighbridges (6 tons) marked with English and foreign weights	No., 1 ,,	60 0 0 ..	60 0 0
2	..	Hand-pumps, with brass barrel, pipes, &c.	,, 2 ,,	30 0 0 ..	60 0 0
..	5	Tanks of cast-iron plates, with planed joints, to hold 500 gallons, pipes, valves, leather hose, &c.	,, 2 ,,	40 0 0 ..	80 0 0
..	2	Sundry tools		60 0 0 ..	60 0 0
9	..	Sundry stores, oil and tallow boxes, rope, chain, spunyarn, padlocks, screw jacks, ironmongery, &c., &c.	,,	200 0 0 ..	200 0 0
13	5	Engines, copper fire-box, brass tubes, steel tyres, with lamps and tools complete	,, 2 ,,	550 0 0 ..	1,100 0 0
1	..	Engine duplicates		100 0 0 ..	100 0 0
4	..	Mineral brake vans, fitted with lockers, &c.	,, 2 ,,	150 0 0 ..	300 0 0
45	..	Open goods wagons, 7 ft. 6 in. × 4 ft. 0 in. × 1 ft. 6 in. ; weight, empty, 16 cwt. ; full, 3 tons	,, 40 ,,	30 0 0 ..	1,200 0 0
4	..	Extra sets of wheels and axles, axle boxes, springs, buffers, &c.	,, 5 ,,	20 0 0 ..	100 0 0
	292	Tons dead weight.			
484		Tons measurement.			

TOTAL COST of all materials and fittings for 10 miles (including sleepers) £7,920 0 0

,, ,, ,, ,, per mile £792 0 0

,, ,, ,, ,, for 20 miles, per mile £750 0 0

ITEMS WHICH MAY OR MAY NOT BE REQUIRED.

					£ s. d.
Cost of plain girders, for 40 ft. span, with rails on top and cross stretchers	per pair	90 0 0
,, ,, 30 ,, ,, ,, ,, ,, ...	,,	50 0 0
,, ,, 20 ,, ,, ,, ,, ,, ...	,,	30 0 0
,, rolled joists 15 ,, ,, ,, ,, ,, ...	,,	18 0 0

K

SIZE No. 4.—(*See Estimate No. 4.*)

Estimate of Materials and Fittings

FOR

TWENTY MILES OF RAILWAY FOR GENERAL TRAFFIC,

WITH

THREE INTERMEDIATE AND TWO TERMINAL STATIONS.

DUTY REQUIRED TO BE DONE.

1. Maximum train load required to be taken at 8 miles an hour on a level = 60 tons.
2.   ,,   ,,   ,,   ,,   5 ,,   ,,   up 1 in 100 = 40 ,,
3.   ,,   ,,   ,,   ,,   3 ,,   ,,   ,, 1 in 35 = 22 ,,
4. Speed with light load = 18 miles an hour.

GAUGE, 2 ft. 0 in. to 3 ft. 6 in.

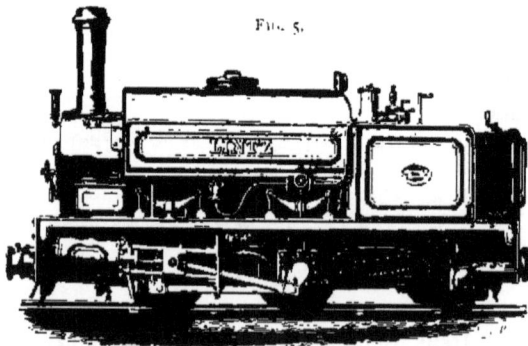

FIG. 5.

DESCRIPTION OF ENGINE SUITABLE FOR THE REQUIREMENTS.

1. Weight in working order, 7 tons, on 6 wheels, 2 ft. diameter, 4 wheels coupled.
2. Cylinders, 6½ in. × 12 in. Heating surface, 142 square feet. Grate area, 3 square feet.
3. Greatest weight on any wheel, 25 cwt. Rails, 20 lbs. per yard. Gauge, 2 ft. 6 in. to 3 ft. 6 in.
4. Fixed wheel-base of Engines when made with six wheels coupled, 7 ft. 5 in.
5. Fixed wheel-base of Engines when made with four wheels coupled, and a " bogie," 3 ft. 6 in.

ESTIMATE No. 4—(for Size No. 4).

Tons for Freight — Measurement.	Tons for Freight — Dead Weight.		Quantity	£ s. d.	£ s. d.
..	667	Rails (20 lbs. per yard), with fish plates	Tons, 667 at	8 0 0 ..	5,336 0 0
..	50	Fish bolts, dog spikes, and fang bolts	„ 50 „	22 0 0 ..	1,100 0 0
802	..	Sleepers, 8 in. × 3 in. × 5 ft. 6 in. (2400 per mile)	No., 48,000 „	0 1 9 ..	4,200 0 0
..	8	Points, steel, 7 ft. tongues, 10 ft. stocks	Sets, 20 „	8 0 0 ..	160 0 0
..	7	Crossings, steel, with check rails	„ 26 „	5 0 0 ..	130 0 0
3	2	Platelayers' tools and trollies	„	130 0 0 ..	130 0 0
2	..	Station signals (semaphore type), with lamps and ladders complete	No., 5 „	18 0 0 ..	90 0 0
4	..	Distant signals, complete with wires, &c.	„ 8 „	25 0 0 ..	200 0 0
2	1	Turntables for turning carriages and wagons .. 8 ft.	„ 1 „	60 0 0 ..	60 0 0
4	6	„ „ tank engines (centre balance) diam. 12 ft.	„ 2 „	120 0 0 ..	240 0 0
2	..	Traversers .. 10 ft.	„ 1 „	55 0 0 ..	55 0 0
4	..	Platform weighing-machines, to weigh 5 cwt.	„ 5 „	8 0 0 ..	40 0 0
8	..	Weighbridges (7 tons), marked with English and foreign weights	„ 2 „	75 0 0 ..	150 0 0
3	..	Hand-pumps, with brass barrels, fly-wheel, pipes, &c.	„ 3 „	30 0 0 ..	90 0 0
..	3	Tanks, of cast-iron plates, with planed joints, supported on columns, with { to hold 500 gals.	„ 1 „	40 0 0 ..	40 0 0
..	7	bracing, pipes, valves, leather hose, &c. } „ 1,000 „	„ 2 „	55 0 0 ..	110 0 0
1	..	Platform water-cranes, 5 in. standard, with leather hose, valve, &c.	„ 1 „	30 0 0 ..	30 0 0
2	..	Station fittings { ticket cases and apparatus, } terminal ..	„ 2 „	120 0 0 ..	240 0 0
		{ lamps, clocks, furniture, &c. } intermediate	„ 3 „	80 0 0 ..	240 0 0
..	3	Sheer legs to lift engines, with 10-ton crab, chains, and tools	„ 1 „	60 0 0 ..	60 0 0
2	3	Goods wharf cranes (3 tons)	„ 2 „	100 0 0 ..	200 0 0
38	..	Workshop fittings, engine, boiler, shafting, lathes and machines, vices, fan and smith's fires, taps and dies, files, chisels, drills, &c.	„ 1,520 0 0 ..		1,520 0 0
8	..	Sundry stores, oil and tallow boxes, rope, chain, spunyarn, padlocks, screw jacks, &c., &c.	„ 350 0 0 ..		350 0 0
76	20	Engines, copper fire-box, brass tubes, steel tyres, with lamps and tools complete	„ 4 „	700 0 0 ..	2,800 0 0
1	..	Engine duplicates		..	140 0 0
12	3	First-class carriages, to hold 12 passengers	„ 3 „	250 0 0 ..	750 0 0
12	3	Second „ „ 15 „	„ 3 „	230 0 0 ..	690 0 0
48	12	Third „ „ 18 „	„ 12 „	210 0 0 ..	2,520 0 0
		(The third-class carriages all fitted with brakes.)			
72	6	Passenger brake vans, fitted for luggage and parcels	„ 3 „	150 0 0 ..	450 0 0
30	..	Low-sided goods wagons, 7 ft. 6 in. × 4 ft. × 1 ft.; weight, empty, 16 cwt.; full, 3½ tons	„ 40 „	28 0 0 ..	1,120 0 0
26	..	High-sided goods wagons, 7 ft. 6 in. × 4 ft. × 2 ft. 6 in. ..	„ 20 „	30 0 0 ..	600 0 0
55	..	Covered goods wagons, 10 ft. × 5 ft. × 5 ft.	„ 15 „	70 0 0 ..	1,150 0 0
22	..	Timber wagons, 10 ft. × 5 ft.	„ 6 „	58 0 0 ..	348 0 0
15	..	Extra sets of wheels and axles, axle boxes, springs, buffers, &c.	„ 15 „	20 0 0 ..	300 0 0
2	..	Extra engine and carriage lamps	„ 15 „	2 10 0 ..	37 10 0
	801	Tons dead weight. TOTAL COST of all materials and fittings for 20 miles ..			£25,576 10 0
1,256		Tons measurement. „ „ „ per mile ..			£1,279 0 0

ITEMS WHICH MAY OR MAY NOT BE REQUIRED.

		£ s. d.
Wire fencing, with standing pillars, galvanized wire, &c.	per mile of road	250 0 0
Level-crossing gates, 12 ft., £8 10s.; 15 ft., £10 10s.		
Electric telegraphs, one wire, with iron poles and insulators	cost per mile	40 0 0
Instruments and batteries for stations	price per station	40 0 0
Strong wrought-iron workshop, with windows and doors complete, say 120 ft. × 35 ft.	price	1,100 0 0
Cost of girder bridges, with main girders, 60 ft. span, with iron cross-girders	each	300 0 0
„ „ „ 50 „ „ „		200 0 0
„ plain girders for 40 ft. span, £110; 30 ft., £60; 20 ft., £40; 15 ft., £30 per pair.		

K 2

SIZE No. 5.—(*See Estimate No.* 5.)

Estimate of Materials and Fittings

FOR

TWENTY MILES OF RAILWAY FOR GENERAL TRAFFIC,

WITH

THREE INTERMEDIATE AND TWO TERMINAL STATIONS.

DUTY REQUIRED TO BE DONE.

1. Maximum train load required to be taken at 9 miles an hour on a level = 120 tons.
2.   ,,   ,,   ,,   ,,   6 ,,   ,,   up 1 in 100 = 60 ,,
3.   ,,   ,,   ,,   ,,   4 ,,   ,,   1 in 35 = 30 ,,
4. Speed with light loads = 20 miles an hour.

GAUGE, 2 ft. 6 in. to 3 ft. 6 in.

FIG. 6.

DESCRIPTION OF ENGINE SUITABLE FOR THE REQUIREMENTS.

1. Weight, in working order, 9 tons, on 6 wheels, 2 ft. 3 in. diameter, all coupled.
2. Cylinders, 8 in. × 12 in. Heating surface, 210 square feet. Grate area, 4 square feet.
3. Greatest weight on any wheel, 30 cwt. Rails, 25 lbs. per yard. Gauge, 2 ft. 6 in. to 3 ft. 6 in.
4. Fixed wheel-base of Engines when made with six wheels coupled, 8 ft. 6 in.
5. Fixed wheel-base of Engines when made with four wheels coupled, with a " bogie," 4 ft. 6 in.

ESTIMATE No. 5—(for Size No. 5).

Tons for Freight.		Item	Quantity	Price £ s. d.	Total £ s. d.
Measurement.	Dead Weight.				
..	834	Rails, 25 lbs. per yard, with fish plates	Tons, 834 at	8 0 0	6,720 0 0
..	60	Fish bolts, dog spikes, and fang bolts	,, 60 ,,	20 0 0	1,200 0 0
1,600	..	Sleepers, 8 in. × 4 in. × 6 ft. (2400 per mile)	No., 48,000 ,,	0 2 3	5,400 0 0
..	9	Points, steel, 9 ft. tongues, 12 ft. stocks	Sets, ,, 20 ,,	9 0 0	180 0 0
..	7	Crossings, steel, with check rails	,, 26 ,,	6 0 0	156 0 0
4	3	Platelayers' tools and trollies	,,	200 0 0	200 0 0
2	..	Station signals (semaphore type), with lamps and ladders complete	No., 5 ,,	18 0 0	90 0 0
4	..	Distant signals, complete with wires, &c.	,, 8 ,,	25 0 0	200 0 0
6	3	Turntables for turning carriages or wagons 8 ft.	,, 3 ,,	60 0 0	180 0 0
4	6	,, ,, tank engines (centre balance) diam. 12 ft.	,, 2 ,,	120 0 0	240 0 0
5	..	Traversers 10 ft.	,, 2 ,,	55 0 0	110 0 0
5	..	Platform weighing-machines, to weigh 5 cwt.	,, 6 ,,	8 0 0	48 0 0
12	..	Weighbridges (7 tons), marked with English and foreign weights	,, 3 ,,	75 0 0	225 0 0
4	..	Hand-pumps, with two brass barrels, fly-wheel, pipes, &c.	,, 4 ,,	40 0 0	160 0 0
2	4	Steam pumps for terminal stations	,, 1 ,,	190 0 0	190 0 0
..	10	Tanks, of cast-iron plates, with planed joints, supported on columns, with bracing, pipes, valves, leather, hose, &c. to hold 2000 gals.	,, 2 ,,	78 0 0	156 0 0
..	15	,, 4000 ,,	,, 2 ,,	120 0 0	240 0 0
2	..	Platform water-cranes, 5 in. standard, with leather hose, valve, &c.	,, 2 ,,	30 0 0	60 0 0
2	..	Station fittings { ticket cases and apparatus, terminal	,, 2 ,,	120 0 0	240 0 0
..	10	lamps, clocks, furniture, &c.} intermediate	,, 3 ,,	80 0 0	240 0 0
		Sheer legs to lift engines, with 15-ton crab, chains, and tools	,, 2 ,,	100 0 0	200 0 0
4	2	Goods warehouse cranes (30 cwt.)	,, 6 ,,	40 0 0	240 0 0
10	20	Goods wharf cranes (5 tons)	,, 4 ,,	180 0 0	720 0 0
20	10	Breakdown cranes (5 tons)	,, 2 ,,	300 0 0	600 0 0
38	..	Workshop fittings, engine, boiler, shafting, lathes and machines, vices, fan and smith's fires, taps and dies, files, chisels, drills, &c.	,, 1,520 0 0		1,520 0 0
8	..	Sundry stores, oil and tallow boxes, rope, chain, spunyarn, padlocks, screw jacks, &c., &c.	,, 400 0 0		400 0 0
80	20	Engines, copper fire-box, brass tubes, steel tyres, with lamps and tools complete	,, 4 ,,	900 0 0	3,600 0 0
1	..	Engine duplicates	,, 180 0 0		180 0 0
18	3	First-class carriages, to hold 16 passengers	,, 3 ,,	280 0 0	840 0 0
21	3	Second ,, ,, 20 ,,	,, 3 ,,	260 0 0	780 0 0
45	9	Third ,, ,, 24 ,,	,, 9 ,,	230 0 0	2,070 0 0
		(The third-class carriages are all fitted with brakes.)			
12	6	Passenger brake vans, fitted for luggage and parcels	,, 3 ,,	150 0 0	450 0 0
50	20	Low-sided goods wagons, 10 ft. × 5 ft. × 1 ft.; weight, empty, full, 5 tons	,, 40 ,,	52 0 0	2,080 0 0
50	10	High-sided goods waggons, 10 ft. × 5 ft. × 3 ft.	,, 20 ,,	55 0 0	1,100 0 0
54	7	Covered goods waggons, 10 ft. × 5 ft. × 5 ft.	,, 15 ,,	73 0 0	1,095 0 0
35	5	Cattle wagons; fitted also for sheep, 10 ft. × 5 ft. 6 in. × 5 ft.	,, 10 ,,	80 0 0	800 0 0
18	3	Horse boxes, 12 ft. × 5 ft. 6 in. × 5 ft. 6 in.	,, 4 ,,	150 0 0	600 0 0
26	..	Timber wagons, 10 ft. × 5 ft.	,, 10 ,,	58 0 0	580 0 0
24	..	Extra sets of wheels and axles, axle boxes, springs, buffers, &c.	,, 20 ,,	24 0 0	480 0 0
4	..	Extra engine and carriage lamps	,, 30 ,,	2 10 0	75 0 0
	1,079	Tons dead weight.	TOTAL COST of all materials and fittings for 20 miles		£34,705 0 0
2,215		Tons measurement.	,, ,, ,, per mile		£1,735 0 0

For a line 40 miles long, the cost per mile will be £1,500.

ITEMS WHICH MAY OR MAY NOT BE REQUIRED.

	£ s. d.
Wire fencing, with straining pillars, galvanized wire, &c. ... per mile of road	250 0 0
Level-crossing gates, 12 ft., £8 10s.; 15 ft., £10 10s.	
Electric telegraphs, one wire, with poles and insulators ... cost per mile	40 0 0
Instruments and batteries for stations ... price per station	40 0 0
Strong wrought-iron workshop, with windows and doors complete, say 150 ft. × 40 ft. price	1,250 0 0
Cost of girder bridges, with main girders 60 ft. span, with iron cross-girders ... each	400 0 0
,, ,, 50 ,,	270 0 0
,, plain girders for 40 ft. span, £130; 30 ft., £60; 20 ft., £40; 15 ft., £30 per pair.	

. **SIZE No. 6.**—(*See Estimate No. 6.*)

Estimate of Materials and Fittings

FOR

TWENTY MILES OF RAILWAY FOR GENERAL TRAFFIC,

WITH

THREE INTERMEDIATE AND TWO TERMINAL STATIONS.

DUTY REQUIRED TO BE DONE.

1. Maximum train load required to be taken at 10 miles an hour on a level = 160 tons.
2.   ,,   ,,   ,,   ,,   7   ,,   ,,   up 1 in 100 = 75 ,,
3.   ,,   ,,   ,,   ,,   4   ,,   ,,   ,, 1 in 35 = 45 ,,
4. Speed with light loads = 22 miles an hour.

GAUGE, 2 ft. 6 in. to 3 ft. 6 in.

FIG. 7.

DESCRIPTION OF ENGINE SUITABLE FOR THE REQUIREMENTS.

1. Weight in working order, 12 tons, on 6 wheels, 2 ft. 6 in. diameter, all coupled.
2. Cylinders 9 in. × 14 in. Heating surface, 300 square feet. Grate area, 5 square feet.
3. Greatest weight on any wheel, 2 tons. Rails, 30 lbs. per yard. Gauge, 2 ft. 6 in. to 3 ft. 6 in.
4. Fixed wheel-base of Engines when made with six wheels coupled, 9 ft. 6 in.
5. Fixed wheel-base of Engines when made with four wheels coupled, and a " bogie," 5 ft. 0 in.

Tons for Freight.		ESTIMATE No. 6—(*for Size No. 6*).			
Measurement.	Dead Weight.			£ s. d.	£ s. d.
..	1,000	Rails, 30 lbs. per yard, with fish plates	Tons, 1,000 at	7 15 0 ..	7,750 0 0
..	80	Fish bolts, dog spikes, and fang bolts	,, 80 ,,	20 0 0 ..	1,600 0 0
1,734	..	Sleepers, 8 in. × 4 in. × 6 ft. 6 in. (2400 per mile) ..	No., 48,000 ,,	0 2 6 ..	3,600 0 0
..	11	Points, steel, 10 ft. tongues, 13 ft. stocks..	Sets, 20 ,,	11 0 0 ..	220 0 0
..	9	Crossings, steel, with check rails	,, 26 ,,	7 0 0 ..	182 0 0
2	2	Platelayers' tools and trollies	,,	100 0 0 ..	100 0 0
2	..	Station signals (semaphore type), with lamps and ladders complete	No., 5 ,,	18 0 0 ..	90 0 0
5	..	Distant signals, complete with wires, &c.	,, 8 ,,	25 0 0 ..	200 0 0
1	1	Point indicators	,, 6 ,,	7 10 0 ..	45 0 0
16	8	Turntables for turning carriages or wagons .. diam. 12 ft.	,, 4 ,,	100 0 0 ..	400 0 0
5	7	,, ,, tank engines (centre balance) ,, 12 ,,	,, 2 ,,	120 0 0 ..	240 0 0
5	..	Traversers .. .. ,, 12 ,,	,, 2 ,,	66 0 0 ..	132 0 0
5	..	Platform weighing-machines to weigh 5 cwt.	,, 6 ,,	8 0 0 ..	48 0 0
27	..	Weighbridges (10 tons), marked with English and foreign weights	,, 3 ,,	100 0 0 ..	300 0 0
..	4	Hand-pumps, with two brass barrels, fly-wheel, pipes, &c. ..	,, 4 ,,	45 0 0 ..	180 0 0
2	4	Steam pumps for terminal stations	,, 1 ,,	190 0 0 ..	190 0 0
..	10	Tanks, of cast-iron plates, with planed joints, supported on columns, with bracing, pipes, valves, leather hose, &c. { to hold 2000 gals.	,, 2 ,,	78 0 0 ..	156 0 0
..	15	,, 4000 ,,	,, 2 ,,	120 0 0 ..	240 0 0
3	..	Water cranes, 5 in. standard, with leather hose, valve, &c. ..	,, 3 ,,	30 0 0 ..	90 0 0
1	..	Hydrants, with hose for washing out boilers	,, 2 ,,	16 0 0 ..	32 0 0
..	17	Water pipes, 4 in. diam.	Yards, 500 ,,	0 8 0 ..	200 0 0
2	..	Station fittings { ticket cases and apparatus, terminal ..	No., 2 ,,	120 0 0 ..	240 0 0
		{ lamps, clocks, furniture, &c. } intermediate	,, 3 ,,	80 0 0 ..	240 0 0
..	12	Sheer legs to lift engines, with 20-ton crab, chains, and tools	,, 2 ,,	130 0 0 ..	260 0 0
4	7	Goods warehouse cranes (30 cwt.)	,, 6 ,,	30 0 0 ..	180 0 0
10	20	Goods wharf cranes (5 tons)	,, 4 ,,	180 0 0 ..	720 0 0
20	10	Breakdown cranes (5 tons)	,, 2 ,,	300 0 0 ..	600 0 0
60	..	Workshop fittings, engine, boiler, shafting, lathes and machines, vices, fan and smith's fires, taps and dies, files, chisels, drills, &c.		,, 2,650 0 0 ..	2,650 0 0
10	..	Sundry stores, oil and tallow boxes, rope, chain, spunyarn, padlocks, screw jacks, &c., &c.		,, 500 0 0 ..	500 0 0
88	24	Engines, copper fire-box, brass tubes, steel tyres, with lamps and tools complete	,, 4 ,, 1,100 0 0 ..		4,400 0 0
2	...	Engine duplicates	,, 250 0 0 ..		250 0 0
18	3	First-class carriages, to hold 16 passengers	,, 3 ,, 300 0 0 ..		900 0 0
21	3	Second ,, ,, 20 ,, ...	,, 3 ,, 275 0 0 ..		825 0 0
60	12	Third ,, ,, 24 ,, ...	,, 12 ,, 250 0 0 ..		3,000 0 0
		(The third-class carriages all fitted with brakes.)			
12	6	Passenger brake vans, fitted for luggage and parcels	,, 3 ,, 160 0 0 ..		480 0 0
50	20	Low-sided goods wagons, 10 ft. × 5 ft. × 1 ft.; weight, empty, 30 cwt.; full, 5 tons	,, 40 ,, 52 0 0 ..		2,080 0 0
50	10	High-sided goods wagons, 10 ft. × 5 ft. × 3 ft.	,, 20 ,, 55 0 0 ..		1,100 0 0
54	7	Covered goods wagons, 10 ft. × 5 ft. × 5 ft.	,, 15 ,, 73 0 0 ..		1,095 0 0
35	5	Cattle wagons, fitted also for sheep, 10 ft. × 5 ft. 6 in. × 5 ft.	,, 10 ,, 80 0 0 ..		800 0 0
18	3	Horse boxes, 12 ft. × 6 in. × 7 ft. 6 in.	,, 4 ,, 150 0 0 ..		600 0 0
26	..	Timber wagons, 10 ft. × 5 ft.	,, 10 ,, 58 0 0 ..		580 0 0
24	..	Extra sets of wheels and axles, axle boxes, springs, buffers, &c.	,, 20 ,, 24 0 0 ..		480 0 0
4	..	Extra engine and carriage lamps	,, 30 ,, 2 10 0 ..		75 0 0
	1,300	Tons dead weight. TOTAL COST of all materials and fittings for 20 miles			£37,850 0 0
2,376		Tons measurement. ,, ,, ,, per mile			£1,892 0 0
		For a line 40 miles long, the cost per mile would be £1,650.			

ITEMS WHICH MAY OR MAY NOT BE REQUIRED.

		£ s. d.
Wire fencing, with straining pillars, galvanized wire, &c.	per mile of road	250 0 0
Level-crossing gates, 12 ft., £8 10s. ; 15 ft., £10 10s.		
Electric telegraphs, one wire, with poles and insulators	cost per mile	40 0 0
Instruments and batteries for stations	price per station	40 0 0
Strong wrought-iron workshop, with windows and doors complete, say 150 ft. × 40 ft.	price	1,250 0 0
Cost of girder bridges, with main girders 60 ft. span, with iron cross-girders 40 ft.	each	500 0 0
,, ,, 50 ,, ,,		350 0 0
,, plain girders for 40 ft. span, £140 ; 30 ft., £80 ; 20 ft., £40 ; 15 ft., £30 per pair.		

(PLATE No. 1.—... Engine No. 6.)

... of ...'s ... fittings

... FOR GENERAL TRAFFIC,

... TERMINAL STATIONS.

... TO BE DONE.

... on a level = 160 tons.
... 1 a 100 = 75 „
... 1 a 35 = 45 „

FIG. 3.

... TION OF ENGINE SUITABLE FOR THE REQUIREM...

... Railway ... 12 tons, on 6 wheels, 2 ft. 6 in. diameter, all con...
... Heating surface, 300 square feet. Grate area,...
... weight on one wheel, 2 tons. Rails, 30 lbs. per yard. Gauge ...
... Engines when made with six wheels coupled, 9 ...
... Engines when made with four wheels coupled,

ESTIMATE No. 7—(for Size No. 7).

Measurement	Dead Weight	Description	Quantity	Rate £ s. d.	Amount £ s. d.
..	2,667	Rails, 40 lbs. per yard, with fish plates	Tons, 2,667 at	7 15 0 ..	20,699 0 0
..	160	Fish bolts, dog spikes, and fang bolts	,, 160 ,,	20 0 0 ..	3,200 0 0
4,388	..	Sleepers 9 in. × 4½ in. × 6 ft. 6 in. (2400 per mile) ..	No., 96,000 ,,	0 2 6 ..	12,000 0 0
..	24	Points, steel, 12 ft. tongues, 15 ft. stocks	Sets, 40 ,,	12 0 0 ..	480 0 0
..	21	Crossings, steel, with check rails	,, 52 ,,	8 0 0 ..	416 0 0
5	4	Platelayer's tools and trollies	,,	250 0 0 ..	250 0 0
3	..	Station signals (semaphore type), with lamps and ladders complete	No., 7 ,,	22 0 0 ..	154 0 0
9	..	Distant signals, complete with wires, &c.	,, 14 ,,	35 0 0 ..	490 0 0
2	2	Point indicators	,, 10 ,,	7 10 0 ..	75 0 0
40	20	Turntables for turning carriages or wagons .. diam. 12 ft.	,, 10 ,,	100 0 0 ..	1,000 0 0
26	13	,, ,, tank engines (centre balance) ,, 18 ft.	,, 3 ,,	200 0 0 ..	600 0 0
36	18	,, ,, engines and tenders .. ,, 42 ft.	,, 2 ,,	350 0 0 ..	700 0 0
10	..	Traversers .. .. 12 ft.	,, 4 ,,	65 0 0 ..	260 0 0
8	..	Platform weighing-machines, to weigh 5 cwt.	,, 9 ,,	8 0 0 ..	72 0 0
57	..	Weighbridge (10 tons), marked with English and foreign weights	,, 7 ,,	100 0 0 ..	700 0 0
4	..	Hand-pumps, with two brass barrels, fly-wheel, pipes, &c. ..	,, 4 ,,	45 0 0 ..	180 0 0
6	12	Steam pumps for terminal stations	,, 2 ,,	320 0 0 ..	640 0 0
..	15	Tanks of cast-iron plates, with planed } to hold 2000 galls.	,, 3 ,,	78 0 0 ..	234 0 0
.:	8	joints, supported on columns with } ,, 4000 ,,	,, 1 ,,	120 0 0 ..	120 0 0
..	9	bracing pipes, valves, leather hose, &c. } ,, 6000 ,,	,, 1 ,,	150 0 0 ..	150 0 0
3	..	Platform water-cranes with swing jib and hose	,, 2 ,,	50 0 0 ..	100 0 0
3	..	Water cranes, 5 in. standard, with leather hose, valve, &c. ..	,, 3 ,,	35 0 0 ..	105 0 0
2	..	Hydrants, with hose for washing out bodies	,, 3 ,,	16 0 0 ..	48 0 0
..	51	Water pipes, 4 in. diameter	Yards, 1,500 ,,	0 8 0 ..	600 0 0
3	..	Station fittings { ticket cases and apparatus, } terminal ..	No., 2 ,,	150 0 0 ..	300 0 0
		{ lamps, clocks, furniture, &c. } intermediate	,, 5 ,,	100 0 0 ..	500 0 0
..	12	Sheer legs to lift engines, with 20-ton crab, chains, and tools	,, 2 ,,	130 0 0 ..	260 0 0
8	14	Goods warehouse cranes (30 cwt.)	,, 12 ,,	40 0 0 ..	480 0 0
20	40	Goods wharf cranes (5 tons)	,, 8 ,,	180 0 0 ..	1,440 0 0
30	15	Breakdown cranes (10 tons)	,, 3 ,,	520 0 0 ..	1,560 0 0
66	..	Workshop fittings, engine, boiler, shafting, lathes and machines, vices, fan and smith's fires, taps and dies, files, chisels, drills, &c.	,, 2,650 0 0 ..		2,650 0 0
17	..	Sundry stores, oil and tallow boxes, rope, chain, spunyarn, padlocks, screw jacks, &c., &c.	,, 800 0 0 ..		800 0 0
180	50	Engines, copper fire-box, brass tubes, steel tyres, with tender lamps and tools complete	,, 7 ,, 1,550 0 0 ..		10,850 0 0
3	1	Engine duplicates	,, 545 0 0 ..		545 0 0
30	5	First-class carriages, to hold 18 passengers	,, 5 ,,	330 0 0 ..	1,650 0 0
30	5	Second ,, ,, 22 ,, ...	,, 5 ,,	300 0 0 ..	1,500 0 0
120	20	Third ,, ,, 26 ,, ...	,, 20 ,,	270 0 0 ..	5,400 0 0
..	..	Brakes on 2nd or 3rd class carriages	,, 10 ,,	15 0 0 ..	150 0 0
25	10	Passenger brake vans, fitted for luggage and parcels ..	,, 5 ,,	170 0 0 ..	850 0 0
216	36	Low-sided goods wagons, 14 ft. × 6 ft. 6 in. × 1 ft.; weight, empty, 50 cwt.; full, 7½ tons	,, 60 ,,	68 0 0 ..	4,280 0 0
120	18	High-sided goods wagons, 14 ft. × 6 ft. 6 in. × 3 ft. ..	,, 30 ,,	73 0 0 ..	2,190 0 0
90	10	Covered goods wagons, 14 ft. × 6 ft. 6 in. × 6 ft.	,, 20 ,,	92 0 0 ..	1,840 0 0
60	6	Cattle wagons, fitted also for sheep, 14 ft. × 6 ft. 6 in. × 6 ft. 6 in.	,, 15 ,,	130 0 0 ..	1,950 0 0
20	5	Horse boxes, 14 ft. × 6 ft. 6 in. × 7 ft.	,, 6 ,,	160 0 0 ..	960 0 0
60	15	Timber wagons, 14 ft. × 5 ft. 6 in.	,, 20 ,,	95 0 0 ..	1,900 0 0
54	..	Extra sets of wheels and axles, axle boxes, springs, buffers, &c.	,, 30 ,,	36 0 0 ..	1,080 0 0
6	..	Extra engine and carriage lamps	,, 40 ,,	2 10 0 ..	100 0 0
5,760	3,306	Tons dead weight. TOTAL COST of all materials and fittings for 40 miles			£86,708 0 0
		Tons measurement. ,, ,, ,, per mile			£2,168 0 0

ITEMS WHICH MAY OR MAY NOT BE REQUIRED.

	£ s. d.
Wire fencing, with straining pillars, galvanized wire, &c. .. .. per mile of road	250 0 0
Level-crossing gates, 12 ft., £8 10s.; 15 ft., £10 10s.	
Electric telegraphs, one wire, with poles and insulators .. .. cost per mile	40 0 0
Instruments and batteries for stations .. .. price per station	40 0 0
Strong wrought-iron workshop, with windows and doors complete, say 150 ft. × 80 ft. price	2,000 0 0
Cost of girder bridges, with main girders 60 ft. span, with iron cross girders .. .. each	600 0 0
,, ,, ,, 50 ,, ,, ,,	430 0 0

,, plain girders for 40 ft. span, £160; 30 ft., £90; 20 ft., £50; 15 ft., £40 per pair.

L

SIZE No. 8.—(*See Estimate No. 8.*)

𝔈stimate of 𝔐aterials anb 𝔉ittings

FOR

FORTY MILES OF RAILWAY FOR GENERAL TRAFFIC,

WITH

FIVE INTERMEDIATE AND TWO TERMINAL STATIONS.

DUTY REQUIRED TO BE DONE.

1. Maximum train load required to be taken at 10 or 12 miles an hour, on a level, = 250 tons.
2. „   „   „   7 or 8   „   up 1 in 100 = 100 „
3. „   „   „   4 or 5   „   up 1 in 35 = 50 „
4. Speed with light loads = 24 miles an hour.

GAUGE, 3 ft. 6 in.

FIG. 10.

DESCRIPTION OF ENGINE SUITABLE FOR THE REQUIREMENTS.

1. Weight in working order, 18 tons, on 6 wheels, 3 ft. 6 in. diameter, all coupled.
2. Cylinders, 10¼ in. by 18 in. Heating surface, 485 square feet. Grate area, 5 square feet.
3. Greatest weight on any wheel, 3½ tons. Rails, 45 lbs. per yard. Gauge, 3 ft. 6 in.
4. Fixed wheel-base of Engines when made with six wheels coupled, 10 ft. 10 in.
5. Fixed wheel-base of Engines when made with four wheels coupled, and a bogie, 6 ft. 6 in.

Tons for Freight.		ESTIMATE No. 8—*(for Size No. 8).*								
Measurement.	Dead Weight.				£	s.	d.	£	s.	d.
..	3,000	Rails, 45 lbs. per yard, with fish plates	Tons, 3,000 at		7	15	0 ..	23,250	0	0
..	180	Fish bolts, dog spikes, and fang bolts	,, 180 ,,		20	0	0 ..	3,600	0	0
4,725	..	Sleepers, 9 in. × 4½ in. × 7 ft. (2400 per mile)	No., 96,000 ,,		0	2	6 ..	12,000	0	0
..	26	Points, steel, 10 ft. tongues, 13 ft. stocks	,, 40 ,,		13	0	0 ..	520	0	0
..	24	Crossings, steel, with check rails	,, 52 ,,		9	0	0 ..	468	0	0
6	5	Platelayers' tools and trollies	,,		300	0	0 ..	300	0	0
3	..	Station signals (semaphore type), with lamps and ladders complete	No., 7 ,,		22	0	0 ..	154	0	0
9	..	Distant signals, complete with wires, &c.	,, 14 ,,		35	0	0 ..	490	0	0
2	2	Point indicators	,, 10 ,,		7	10	0 ..	75	0	0
40	20	Turntables for turning carriages or wagons .. diam. 12 ft.	,, 10 ,,		100	0	0 ..	1,000	0	0
26	13	,, ,, tank engines (centre balance) ,, 18 ,,	,, 3 ,,		200	0	0 ..	600	0	0
36	18	,, ,, engines and tender .. .. ,, 42 ,,	,, 2 ,,		350	0	0 ..	700	0	0
10	..	Traversers .. .. ,, 12 ,,	,, 4 ,,		65	0	0 ..	260	0	0
8	..	Platform weighing machines, to weigh 5 cwt.	,, 9 ,,		8	0	0 ..	72	0	0
57	..	Weighbridge (10 tons) marked with English and foreign weights	,, 7 ,,		100	0	0 ..	700	0	0
4	..	Hand-pumps, with two brass barrels, fly-wheel, pipes, &c. ..	,, 4 ,,		45	0	0 ..	180	0	0
6	12	Steam pumps for terminal stations	,, 2 ,,		320	0	0 ..	640	0	0
..	15	Tanks, of cast-iron plates, with planed } to hold 2000 gals.	,, 3 ,,		78	0	0 ..	234	0	0
..	8	joints, supported on columns, with } ,, 4000 ,,	,, 1 ,,		120	0	0 ..	120	0	0
..	9	bracing, pipes, valves, leather hose, &c. } ,, 6000 ,,	,, 1 ,,		150	0	0 ..	150	0	0
3	..	Platform water-cranes, with swing jib and hose	,, 2 ,,		50	0	0 ..	100	0	0
3	..	Water cranes, 5 in. standard, with leather hose, valve, &c. ..	,, 3 ,,		35	0	0 ..	105	0	0
1	..	Hydrants, with hose for washing out boilers	,, 2 ,,		16	0	0 ..	32	0	0
..	51	Water pipes, 4 in. diameter	Yards, 500 ,,		0	8	0 ..	600	0	0
		Station fittings { ticket cases and apparatus, } terminal	No., 2 ,,		150	0	0 ..	300	0	0
3	..	{ lamps, clocks, furniture, &c. } intermediate	,, 5 ,,		100	0	0 ..	500	0	0
..	14	Sheer legs, to lift engines, with 25-ton crab, chains, and tools	,, 2 ,,		160	0	0 ..	320	0	0
30	24	Goods warehouse cranes (30 cwt.)	,, 20 ,,		40	0	0 ..	800	0	0
20	40	Goods wharf cranes (5 tons)	,, 8 ,,		180	0	0 ..	1,440	0	0
30	15	Breakdown cranes (10 tons)	,, 3 ,,		520	0	0 ..	1,560	0	0
95	..	Workshop fittings, engine, boiler, shafting, lathes and machines, vices, fan and smith's fires, taps and dies, files, chisels, drills, &c.	,, 4,200		0	0	..	4,200	0	0
21	..	Sundry stores, oil and tallow boxes, rope, chain, spun-yarn, padlocks, screw jacks, &c., &c.	,, 1,000		0	0	..	1,000	0	0
210	80	Engines, copper fire-box, brass tubes, steel tyres, with lamps and tools complete	,, 8 ,,		1,500	0	0 ..	12,000	0	0
3	1	Engine duplicates	,,		600	0	0 ..	600	0	0
30	5	First class carriages, to hold 18 passengers	,, 5 ,,		330	0	0 ..	1,650	0	0
30	5	Second ,, ,, 22 ,, ...	,, 5 ,,		300	0	0 ..	1,500	0	0
120	20	Third ,, ,, 26 ,, ...	,, 20 ,,		270	0	0 ..	5,400	0	0
..	..	Brakes on 2nd or 3rd class carriages	,, 10 ,,		15	0	0 ..	150	0	0
25	10	Passenger brake vans, fitted for luggage and parcels	,, 5 ,,		170	0	0 ..	850	0	0
300	48	Low-sided goods wagons, 14 ft. × 6 ft. 6 in. × 1 ft. ; weight, empty, 50 cwt. ; full, 7½ tons	,, 80 ,,		68	0	0 ..	5,440	0	0
160	24	High-sided goods wagons, 14 ft. × 6 ft. 6 in. × 3 ft. ..	,, 40 ,,		73	0	0 ..	2,920	0	0
180	20	Covered goods wagons, 14 ft. × 6 ft. 6 in. × 6 ft.	,, 40 ,,		92	0	0 ..	3,680	0	0
80	8	Cattle wagons, fitted also for sheep, 14 ft. × 6 ft. 6 in. × 6 ft. 6 in.	,, 20 ,,		180	0	0 ..	3,600	0	0
34	17	Horse boxes, 14 ft. × 6 ft. 6 in. × 7 ft. 6 in.	,, 10 ,,		160	0	0 ..	1,600	0	0
60	15	Timber wagons, 14 ft. × 5 ft. 6 in.	,, 20 ,,		95	0	0 ..	1,900	0	0
54	..	Extra sets of wheels and axles, axle boxes, springs, buffers, &c.	Sets, 30 ,,		36	0	0 ..	1,080	0	0
6	..	Extra engine and carriage lamps	,, 40 ,,		2	10	0 ..	100	0	0
	3,729	Tons dead weight.								
6,430		Tons measurement.	TOTAL COST of all materials and fittings for 40 miles £98,840					0	0	
			,, ,, ,, per mile £2,471					0	0	

ITEMS WHICH MAY OR MAY NOT BE REQUIRED.

	£	s.	d.
Wire fencing, with straining pillars, galvanised wire, &c. .. .. per mile of road	250	0	0
Level-crossing gates, 12 ft., £8 10s. ; 15 ft. £10 10s.			
Electric telegraphs, one wire, with poles and insulators .. .. cost per mile	40	0	0
Instruments and batteries for stations .. .. price per station	40	0	0
Strong wrought-iron workshop, with windows and doors complete, say 150 ft. × 80 ft. price	2,000	0	0
Cost of girder bridges, with main girders 60 ft. span, with iron cross-girders .. .. each	700	0	0
,, ,, ,, 50 ,, ... ,,	510	0	0
,, plain girders for 40 ft. span, £180 ; 30 ft., £100 ; 20 ft. £50 ; 15 ft., £40 per pair.			

L 2

APPENDIX I.

MACHINERY FOR CUTTING SLEEPERS.

(Manufactured by Messrs. A. Ransome & Co., Chelsea, London.)

In countries where timber is plentiful it often happens that sleepers can be cut close alongside the railway. To meet such cases estimates are subjoined of machines suitable for this purpose.

For cutting out small sleepers, a plain saw bench, as shown in Fig. 42, will suffice to cut out 500 sleepers per day, sawn on two sides, or 300 if sawn on four sides. The framing is cast in one piece, by

Fig. 42.

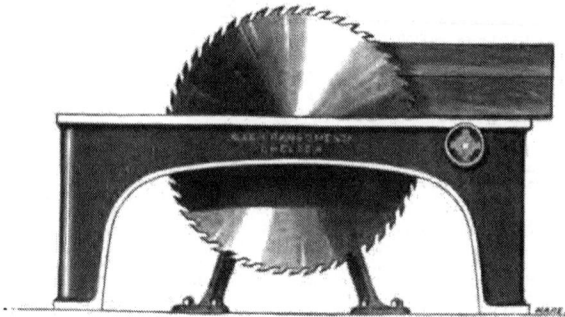

SLEEPER SAW BENCH.

which great strength and stiffness are attained. The bearings in which the saw spindles run are of great length, and are fitted with special lubricating arrangements, and in consequence they last for many years in constant work. Care should be taken to keep the hemp packing well tucked in, so as to ensure a steady pressure sufficient to warm the saw all over alike, and thus cause it to expand equally and work smoothly.

Cost of bench with 36 inch saw..	..	..	..	£39
Cost of six-horse power portable engine to drive it	..			225
Spare saws and spare parts for the engine	..		..	25

For heavy work it is better to have a self-acting saw bench, as shown in Fig. 43. In this machine the feed motion is driven from the saw spindle, and can be varied so as to bring the timber forward at rates varying from 15 to 60 feet a minute. This arrangement requires a little more steam power than a plain saw bench does, but it saves the heavy manual labour of pushing the logs forward, and it does nearly double the work of the ordinary bench.

FIG. 43.

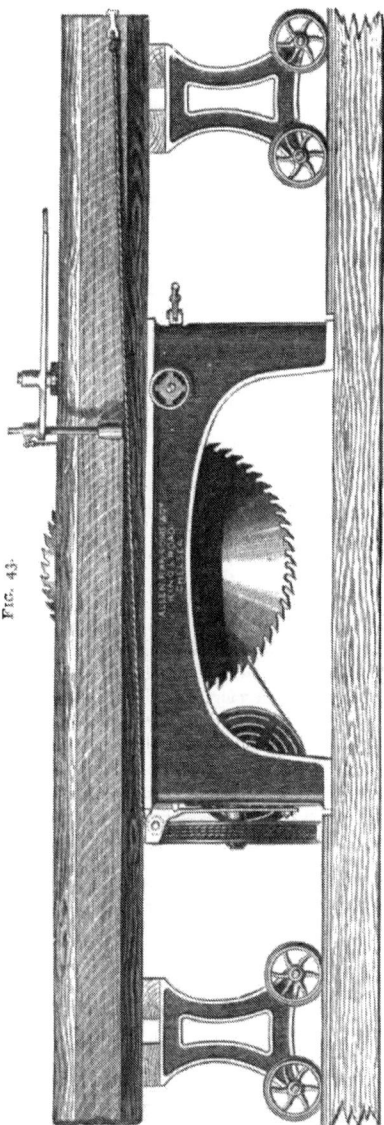

SELF-ACTING SAW BENCH.

FIG. 44.

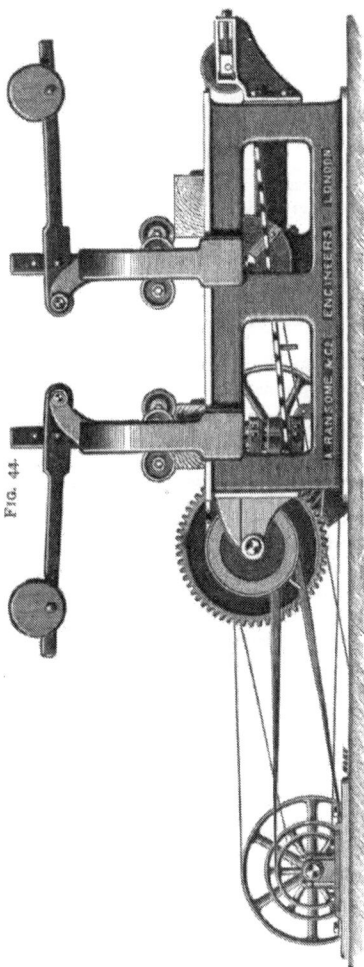

MACHINE FOR ADZING AND BORING SLEEPERS.

Cost of bench with 36 inch saw ..	..	..	..	£58
„ Timber carriages and rails	..	..	..	18
„ 8-horse power portable engine ..	..	..	260
Spare saws and spare parts for the engine	..	..	30

When there is much work to be done, and the timber is hard, a machine for adzing and boring will be found very serviceable. (Fig. 44.)

When once adjusted to suit the required gauge, all the rail seatings are cut so exactly alike, that, in laying the road, the gauge is determined for certain at every sleeper; and the road can thus be laid with greater expedition and with cheaper labour.

The machine takes the sleepers from the saw, and finishes them ready for laying. It adzes the rail seatings and bores the four spike holes in one operation, finishing the sleepers at the rate of two a minute.

When placed upon the machine the sleepers are immediately caught by a pair of wrought-iron dogs fixed to two endless chains, which carry them first over the adzes, thus forming the two seatings simultaneously, and then over the augers, which bore the four holes all at once.

The machine is entirely self-acting, and, when once set for work, all that is required is to lay the rough sleepers on at one end, and remove them as they leave it finished at the other.

The seatings are formed by cutters fixed to a pair of strong adze blocks, which adze the sleeper as it passes over them, and form perfectly true and level beds for the rails. These cutters may be set at any angle, so as to give any required cant to the rail.

One of the principal features in this machine is the self-acting arrangement for stopping the sleepers when in precisely the right position over the augers, and again moving them forward after they are bored. This is done as follows :—The two tumblers which work the endless chains run loose upon a horizontal shaft, and are each provided with clutch teeth, into which work the corresponding clutches, sliding upon feathers upon the same shaft. The clutches are so arranged that as they revolve they disengage themselves from the chain tumbler during part of each revolution, thus causing it to stop during that time. A cam is fixed in such a position upon the driving shaft that during the time the clutches are disengaged from the tumbler, and consequently, when the sleeper is at rest, it acts upon a lever, which raises the augers, and thus simultaneously bores the four holes. The cam having passed the lever, the augers fall clear of the sleeper, the clutches at once come into contact with the tumbler, and the sleepers are again moved forward.

The machine with its counter shaft may be fixed upon a strong frame of timber bedded into the ground, and can be driven direct from any portable engine of not less than 6-horse power, or it may be driven by the same engine which drives the saw bench, in which case an engine of 10-horse power would be required to drive the two machines.

Cost of sleeper adzing and boring machine for sleepers up to 7 feet long ..	..	..	..	..	..	..	..	£180
Cost of 10-horse power portable engine	..	..	..	..	305
„ spare parts for engine and machine	..	..	..	45

The machine shown in the woodcut is for adzing and boring rectangular sleepers. A machine for treating half-round sleepers is made at about the same cost.

FIG. 45.

MACHINES FOR FELLING TREES BY STEAM POWER.

A proposal to use sleepers cut on the spot is of course suggested by the economy expected to result. To secure this economy it is very necessary to have the best class of plant for the purpose, and care should be taken to have adequate driving power. A six or eight-horse power engine will drive a saw bench very well if good fuel is to be had, but if it is intended to use the waste wood as fuel, an engine and boiler of double the size should be provided. The economy of working with refuse fuel will of course cover the additional cost of the engine many times over ; but if it be attempted to secure the advantage without the necessary outlay, of course delay and disappointment must be the result. Well furnished plant, with ample steam power, is sure to give the best result.

Where the sleepers are to be cut from trees of ordinary sleeper sizes the trees can be easily felled in the ordinary way. As, however, there are many countries where the trees are large, and involve great labour in felling, a description is subjoined of a machine for this purpose.

Ransome's Patent Tree Feller consists of a steam cylinder of small diameter, having a long stroke, attached to a light cast-iron bed plate, upon which it is arranged to pivot on its centre, the pivoting motion being readily worked by a lever, as shown in the engraving. The saw is fixed direct to the end of the piston rod, which is caused to travel in a true line when at work by guides, and the range of the pivoting motion of the cylinder is such as to enable the saw to pass through the largest logs that are ordinarily to be met with, without moving the bed plate. A strong wrought-iron strut is attached to the bed plate, and this is furnished with two fangs, which are made to bite into the butt of the tree by a chain passed round it just below the saw cut, and drawn taut by a powerful screw.

The machine is supplied with steam at a high pressure from a small portable boiler through a strong flexible steam pipe, and as this may be of considerable length, the boiler may remain in one place, until the machine has cut down all the trees, within a radius, which is determined by the length of the steam pipe.

From the foregoing description it will be clearly seen, that the only fixing the machine requires after it is laid down against the tree, is to draw it tight against the butt by the chain and screw above referred to; and as the whole apparatus, exclusive of the boiler, does not weigh more than about 3 cwt., it is readily carried about slung on poles between four men. The steam pipe does not require to be disconnected while the machine is being removed, and a special valve is attached by which it can be instantly started at any part of the stroke.

As the pressure of steam is high, the machine works with great rapidity, and under ordinary circumstances it will fell from four to six trees, averaging 30 inches in diameter, in an hour ; and as it cuts nearly close to the ground, it saves a considerable amount of timber which is lost when felling with the axe.

As the machine will work in any position, it will fell trees growing on slopes or in hedgerows, and by a simple apparatus easily attached, it can be fixed so as to crosscut logs to length when lying on the ground.

Size No. 1, to fell trees up to 24 inch diameter	..	..	..	..	£60	0	0
„ No. 2,	„	36 inch	„	..	..	..	75	0	0
„ No. 3,	„	48 inch	„	..	..	..	90	0	0

Boilers for above, with wheels and shafts complete—

Size for No. 1 Tree Feller	..	..	..	..	..	..	..	85	0	0
„ „ 2 „	..	..	..	..	..	..	..	100	0	0
„ „ 3 „	..	..	..	..	..	..	..	120	0	0
Cost of spare saws and fittings	..	..	..	..	..	..	5	0	0

COST OF COMPLETE APPARATUS.

	£	s	d
No. 2 Tree Feller, to fell trees up to 36 inch diameter	75	0	0
Boiler for ditto	100	0	0
Spare parts	5	0	0
Self-acting saw bench with 36 inch saw	58	0	0
Spare saws	10	0	0
Timber carriages and rails	18	0	0
Timber jim, crow bars, and sundries	28	0	0
Sleeper adzing and boring machine	180	0	0
10-horse power portable engine	305	0	0
Spare parts for engine and machine	45	0	0
	£824	0	0
Packing, freight and insurance, say	120	0	0
Total	£944	0	0

If sleepers are required for 20 miles of railway, there would be 48,000 required ; and, if the whole cost of the above plant be charged to that quantity, it would represent an average of 4½d. per sleeper ; to which if we add for labour and fuel 2½d. per sleeper, we have 7d. per sleeper as the cost over and above any charges there might be payable for the standing trees. Seeing that sleepers from England cost from 1s. 3d. to 2s. 6d. each, according to size, and that to this an average freight charge of 50 per cent. must be added, it is clear that a considerable saving may be made whenever timber is to be had on the spot.

The above plant would cut out and finish sleepers for 20 miles of railway in about 10 weeks.

If the trees can be felled a few months in advance of the time for cutting them up, so much the better.

APPENDIX II.

WORKSHOP MACHINERY AND TOOLS SUITABLE FOR ESTIMATES, Nos. 1, 2, & 3.

	£	s.	d.
Six horse power (nominal) horizontal engine with multitubular boiler	250	0	0
Thirty feet run of 2¼ inch shafting, with bearings and wall brackets	24	0	0
One 12 inch lathe to turn wheels on their axles, fitted to bore cylinders, and for general purposes	185	0	0
One 7 inch lathe 4 feet centres, self-acting and screw-cutting	70	0	0
One radial drilling machine, with table, drills, and tools	110	0	0
One slotting machine 6 inch, self-acting	75	0	0
One planing machine 3 feet stroke by 16 by 16	60	0	0
One set taps and dies ⅜ inch to 1¼ inch	15	0	0
One set gas pipe taps and dies ⅜ inch to 1¼ inch (inside bore)	15	0	0
Two sets standard rhymers ⅜ inch to 1¼ inch	8	0	0
Three vices, with files and chisels, &c.	10	0	0
Two portable forges	20	0	0
Two smiths' fires made of wrought iron (with hoods)	30	0	0
Two full sets smiths' tools and anvils	36	0	0
One fan with driving drums, counter-shaft bearings, pipes, &c.	50	0	0
One hundred and sixty feet of leather belting, assorted sizes	10	0	0
Two cwt. tool steel, assorted	6	0	0
One boiler prover	20	0	0
One hydraulic for putting wheels on to axles and taking them off	40	0	0
One 3 feet saw bench	36	0	0
One grindstone and trough with pulleys, &c.	20	0	0
Sundries	40	0	0
	£1,130	0	0

APPENDIX II.—*continued.*

WORKSHOP MACHINERY AND TOOLS SUITABLE FOR ESTIMATES, Nos. 4 & 5.

	£	s.	d.
Ten horse power (nominal) horizontal engine with multitublar boiler	350	0	0
Forty feet run 2½ inch shafting, with bearings and wall brackets	32	0	0
One 16 inch lathe to turn wheels on their axles, fitted to bore cylinders, and for general purposes, 10 feet between centres	250	0	0
One 7 inch lathe 4 feet centres, self-acting and screw-cutting	70	0	0
One radial drilling machine, with table, drills, and tools	110	0	0
One bench drilling machine, with fast and loose pulley	20	0	0
One slotting machine 6 inch self-acting	75	0	0
One planing machine 4 feet stroke, by 2 feet by 2 feet	90	0	0
One screwing machine (patent American)	60	0	0
One shearing and punching machine, to punch ¾ inch holes in ¾ inch plates	90	0	0
One set taps and dies ¼ inch to 1½ inch	15	0	0
One set gaspipe taps and dies ¼ inch to 1½ inch inside bore	15	0	0
Four vices, with files and chisels, &c.	20	0	0
Two smiths' fires, made of wrought iron (with hoods)	30	0	0
Two sets smiths' tools and anvils	40	0	0
One fan with driving drums, counter-shaft bearings, pipes, &c.	50	0	0
Two hundred and forty feet of leather belting, assorted sizes	15	0	0
Four cwt. of tool steel, assorted	12	0	0
One boiler prover	20	0	0
One hydraulic for putting wheels on and taking off	40	0	0
One grindstone and trough, with pulleys	20	0	0
One saw bench, 36 inch	36	0	0
Hand braces, ratchet braces, rhymers, surface plates, hand vices, cramps, and other sundries	60	0	0
	£1,520	0	0

M 2

APPENDIX II.—*continued.*

WORKSHOP MACHINERY AND TOOLS SUITABLE FOR ESTIMATES, Nos. 6 & 7.

	£	s.	d.
Fifteen horse power (nominal) horizontal engine, with multitubular boiler	475	0	0
Sixty feet run of 3 inch shafting, with bearings and wall brackets	60	0	0
One wheel lathe to turn wheels on their axles	400	0	0
One 10 inch lathe 5 feet centres, self-acting and screw-cutting	120	0	0
One 7 inch lathe 4 feet centres, self-acting and screw-cutting	70	0	0
One boring mill for boring cylinders	150	0	0
One radial drilling machine, with tables, drills and tools	110	0	0
One bench drilling machine, with fast and loose pulley	20	0	0
One slotting machine, 8 inch, self-acting	90	0	0
One shaping machine, 6 inch, self-acting	85	0	0
One planing machine, 4 feet stroke by 2 feet by 2 feet	90	0	0
One screwing machine (patent American)	60	0	0
One shearing and punching machine to punch ½-inch holes in ½-inch plate	130	0	0
One set taps and dies ¼ inch to 1¼ inch	15	0	0
One set gas pipe taps and dies, ½ inch to 1½ inch (inside bore)	15	0	0
Six vices, with files and chisels, &c.	28	0	0
Four smiths' fires, made of wrought iron (with hoods)	60	0	0
Three portable forges	30	0	0
Four sets smiths' tools and anvils	60	0	0
One steam hammer, 6 cwt.	110	0	0
One fan, with driving drums, counter-shaft bearings, pipes, &c.	50	0	0
One hydraulic, for putting wheels on to their axles and taking them off	40	0	0
One hydraulic press for smiths' use	60	0	0
One overhead travelling crane, 6 tons, 38 feet span	120	0	0
Three hundred feet of leather belting, assorted sizes	20	0	0
Eight cwt. tool steel, assorted	24	0	0
One grindstone and trough, with pulleys	20	0	0
One 3 feet saw bench	36	0	0
Hand braces, ratchet braces, rhymers, surface plates, hand-vices, cramps, and other sundries	102	0	0
	£2,650	0	0

APPENDIX II.—*continued.*

WORKSHOP MACHINERY AND TOOLS SUITABLE FOR ESTIMATE, No. 8.

	£	s.	d.
Twenty horse power (nominal) horizontal engine, with multitubular boiler	600	0	0
Ninety feet run of 3¼ inch shafting, with bearings and wall brackets	100	0	0
One wheel lathe to turn wheels on their axles	400	0	0
One 10 inch lathe, 5 feet centres, self-acting and screw-cutting	120	0	0
Two 7 inch lathes, 4 feet centres, self-acting and screw-cutting	140	0	0
One boring mill for boring cylinders	230	0	0
One radial drilling machine, with table, drills, and tools	120	0	0
One standard drilling machine, with swing table	50	0	0
One bench drilling machine, with fast-and-loose pulley	20	0	0
One slotting machine, 12 inch, self-acting	105	0	0
One shaping machine, 6 inch stroke	85	0	0
One planing machine, 6 feet stroke by 2 feet 6 inches by 2 feet 6 inches	110	0	0
One screwing machine (patent American)	55	0	0
One shearing and punching machine to punch ¾ inch holes in ¾ inch plates	130	0	0
Two sets taps and dies, ¼ inch to 1¼ inch	30	0	0
One set gaspipe taps and dies, ¼ inch to 1¼ inch (inside bore)	15	0	0
Ten vices, with files and chisels, &c.	40	0	0
Eight smiths' fires, made of wrought iron (with hoods)	120	0	0
Four portable forges	40	0	0
Six sets smiths' anvils and tools	70	0	0
One steam hammer, 10 cwt.	170	0	0
One fan, with driving drums, counter-shaft, bearings, pipes, &c.	70	0	0
One cupola to melt 4 tons per hour	100	0	0
Six foundry ladles, assorted	15	0	0
One plate or wheel tyre furnace, fittings for	70	0	0
One wheel tyre blocking apparatus, with cooling tank	50	0	0
One wheel tyre crane, 1¼ tons	45	0	0
One foundry crane	95	0	0
One overhead travelling crane, 10 tons, 38 feet span	300	0	0
One hydraulic press for putting wheels on to their axles and taking them off	40	0	0
One hydraulic press for smiths' use	60	0	0
Four hundred feet of leather belting, assorted sizes	35	0	0
Twelve cwt. tool steel, assorted	36	0	0
One boiler prover	20	0	0
One general joiner machine	245	0	0
One 36 inch saw bench	36	0	0
One band saw machine	65	0	0
Two grindstones and trough	35	0	0
Hand braces, ratchet braces, rhymers, surface plates, hand vices, cramps, tubing tools, and other sundries	143	0	0
	£4,200	0	0

APPENDIX III.

CLEMINSON'S SYSTEM OF RADIAL AXLES.

This invention is quite recent, but it has in a space of a few months attained very marked success.

Whilst the system is applicable to all railways, it is of especial advantage to the class of railways advocated in the present volume.

Sharp curves afford facilities for cheapness of construction, but they greatly increase the cost of haulage of trains, when the vehicles are of ordinary construction. Considerable mitigation of the resistance of vehicles on curves is obtained by the use of ordinary bogies, but this is gained by a very serious increase in the weight of vehicles, as well by the addition of the bogies themselves, as by the extra strength of framing required.

In Mr. Cleminson's system on the contrary, the carriage body and the underframe continue precisely as at present. Three pairs of wheels are used with axle boxes, axle guards, and springs, precisely as in ordinary stock. Each pair of axle guards and springs are fixed to a very light bogie frame, the four corners of which are faced with planed, self-lubricating surfaces about 4" × 6" and upon these the underframe of the carriage (also provided with similar planed surfaces) rests, as shown

FIG. 46.

CARRIAGE WITH MR. CLEMINSON'S FLEXIBLE WHEEL-BASE.

in Fig. 46. These sliding surfaces are immediately over each end of each spring, and thus the underframe sits at all times on both ends of all springs in the most direct manner. Thus there is no transmission of weight through the bogie frames, and they may therefore be of very light construction.

The two end frames are pivoted in their centres to the underframe of the carriage, but no part of the weight rests on the pivots. The whole weight is taken directly by the springs alone ; it is obvious that this method of suspending the vehicle has many great advantages in passing over irregularities of the road. The central bogie is not pivoted, but is arranged so that it can slide laterally in guides under the underframe. The three bogies are articulated to one another in such a way, that when running on a curve, every axle becomes truly radial, as shown in Fig. 47, and thus the vehicle passes round the curve with almost the same freedom as on a straight line. The results already attained on five of the most important railways in England are so satisfactory, that on one of them, a new suite of saloon carriages for Her Majesty's use is being constructed on this principle. Bearing in mind the cautious way in which all new inventions are received in England, this seems to be no trifling testimony to the ascertained capabilities of the one under notice.

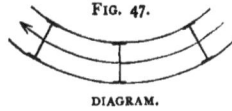

FIG. 47.

DIAGRAM.

The features which particularly commend it in connection with the subject of light railways, are its lightness, cheapness, and perfection of its action. It adds scarcely anything to the cost of a vehicle of ordinary dimensions, and it enables vehicles to be made of greater length, at a considerable reduction in proportion of cost and of dead weight to paying load.

FIG. 48.

WAGON WITH FLEXIBLE WHEEL-BASE.

Fig. 48 is an illustration of a long wagon for the purpose of carrying rails, timber, &c., &c. On all railways the carrying of rails for example is a necessity, and a scheme of light railways could scarcely be said to be complete without some provision of the kind. Not only for vehicles which must be of extra length for such purposes, but also for ordinary vehicles of every kind, this plan is worthy of consideration. It would clearly be a mistake to argue that because an engine with six wheels coupled must have a rigid wheel-base of eight or ten feet, therefore all the other vehicles may just as well have the same rigid wheel-base. There can be no doubt that on a line with many curves a train of given weight fitted with radial axles can be hauled with a less expenditure of fuel than a train not so fitted. Moreover, if the ruling gradient of a length of a railway happens to be on a curve, a simple method of neutralizing the curve will enable the train load to be considerably increased.

The additional cost of this system to ordinary stock is about 5 per cent., and this will be found to be money well spent, by reason of the reduced wear and tear, reduced cost of haulage per ton of load, and practicable increase of train loads.

It has been very often severely tested, at speeds of sixty-five and even seventy miles an hour, on some of the worst curves of the London, Chatham, and Dover Railway, and of the London and South-

Western Railway. A large saloon carriage, with 23 feet wheel-base, passes round curves of 7 chains radius at high speeds with ease and steadiness, and consequently with perfect comfort to the occupants.

Another striking instance of its capabilities may be found on the North Wales narrow-gauge railways : this line is an almost interminable series of curves and reverse curves of radii varying from 2 chains upwards ; the Cleminson stock working over these has a wheel-base of 23 feet, and though it is frequently run at twenty miles an hour, there is a steadiness of motion, ease of traction, and absence of oscillation unapproachable by either the bogie or rigid wheel-base stock on the line.

The invention is now fairly incorporated into the general practice of several leading railway companies. Its importance can scarcely be overrated, seeing that it now enables railways to be made with such curves as to make it quite practicable to go round the hills, instead of going through them by tunnels, or over them by heavy gradients, and this, too, without involving the use of such heavy stock as the ordinary " bogie " system necessitates.

———•◦;◉;◦•———

THE FIRST RAILWAY IN CHINA.

SHANGHAI AND WOOSUNG.

The introduction of Western improvements of steam travelling into China has occupied the attention of practical philosophers during the last thirty years. Steamboats were introduced into Chinese waters twenty years ago; and with such success, that many steamers were soon traversing the rivers under Chinese ownership and direction.

For twenty years or more the introduction of railways into China has been a favourite topic. In the year 1863–64 Sir R. Macdonald Stephenson, who had already done so much for the development of railways in India, visited China with a like object. He was received with the utmost enthusiasm by the Chinese people, and it certainly seemed that railways must soon be introduced. Sir Macdonald Stephenson pointed out the salient features so distinctly that they deserve to be recorded as the basis of what he proposed.

1. That China was so deficient in roads, and the canals were so constantly out of repair, that she needed railways more than any other country, and would benefit more by their introduction than any other country had done.

2. That the physical features of the country were highly favourable for the construction of railways.

3. That materials of most kinds, and labour of all kinds, were to be had in the country, and that Chinese railways might be made by the Chinese themselves.

4. That any financial arrangements required could be easily made by English houses, inasmuch as the railways he proposed were certain to be remunerative.

5. Local Governors have, however, opposed railways, under the idea that they would interfere with their privileges in turning the local taxing powers to best account. Sir Macdonald Stephenson, therefore, pointed out that railways would afford the best possible facilities to the Chinese Government and authorities for collecting all legitimate taxes.

6. That railways would not wholly supersede canal traffic, and that they would give much more employment than they would take away.

These leading features having been fully recognized and set forth, one would have naturally supposed that some result might have been hoped for. But the Empire of China is like no other. Descended from a remote antiquity, her rulers have ever held the doctrine, that she is the hereditary fountain of all knowledge and advancement; and that any suggestions coming from the West must be, in the nature of things, retrogressive.

They themselves have a proverb to the effect that "true wisdom lies not in a servile imitation of the past, but in availing ourselves of such improvements as experience may have suggested." Words of wisdom which, it is evident, China does not yet fully comprehend.

Sir Macdonald Stephenson visited China on public grounds, and he therefore, very consistently, proposed an extensive scheme of railways for that country, being anxious to guard against the mistakes and consequent extravagances which had occurred in the construction of railways in other countries. The completeness with which he placed the subject before the Chinese authorities was probably the very reason why nothing was done. Possibly if Sir Macdonald Stephenson had proposed only some one line, a trial might have been made; but as his proposals embraced a scheme for bringing Calcutta into railway communication with Pekin, it is more than likely that this led to his intentions being misunderstood.

Having in view the non-success of Sir Macdonald Stephenson's extensive scheme, it now seemed to some of those interested, that possibly a short line, at one or other of the ports, which are more amenable to European influences, would be more likely to attain success.

Therefore, in the year 1865, a Company was proposed for constructing a railway from Shanghai to Woosung, with a jetty and bonded warehouses at the latter place, by which the necessity for the passage of the larger class of steamers, up the difficult and changeable navigation to Shanghai, could be avoided. Mr. Henry Robinson, M. Inst. C.E., was the engineer of this Company, and he planned a line of railway, over a similar route to that of the one recently constructed. He proposed to carry it on piles and girders where it might interfere with graves, or otherwise disturb vested interests or prejudices of the people.

Mr. Pickwood, for many years resident at Shanghai, was actively interested in this project, and devoted much time, both in China and in England, to maturing the various arrangements connected with the undertaking. Most unfortunately, his premature death deprived the project of its chief promoter.

In all the various efforts for introducing railways into China, the house of Messrs. Jardine Matheson and Co. had always given cordial support; but the disappointments and delays, again and again incurred, led them eventually to the opinion that the only way to make a railway would be to quietly acquire the necessary land, and make the line as an undertaking of their own, under the sole control of themselves and their friends.

In all the European settlements in China the construction of roads has to be undertaken by the Europeans, and in many districts these roads are the only ones worthy of the name; this practice has in some instances led to roads being constructed by Europeans even beyond the limits of their settlements.

An expansion of this practice was Messrs. Jardine Matheson and Co's. proposal to construct a road (a road of some sort) from Shanghai to Woosung. The acquisition of land for this purpose was, of course, a work of great difficulty, and one which could only be carried to success by being conducted in the most quiet and patient manner, and therefore much time was necessary for the preliminary stages.

In the year 1872 the author (ignorant of the project which Messrs. Jardine Matheson and Co. had in hand) conceived the idea that a step might be gained by sending some engines, carriages, and rails to the Emperor of China, on the occasion of his majority and marriage. The notion was that if the Emperor once experienced the pleasure of a ten-miles railway ride, his first feeling would probably be one of impatience at the shortness of the run, and that an extension of the line would be speedily ordered.

This project took a definite shape in 1873. The Duke of Sutherland kindly lent his powerful aid to obtain for it a good hearing, and several meetings were held at Stafford House under the presidency of His Grace. The King of the Belgians also took a most lively interest in the proposal, and the author was honoured with several conversations with His Majesty on the subject—conversations marked by the enlightened views which His Majesty is so well known to entertain. British and Belgian statesmen warmly supported the proposal, and there seemed to be no doubt as to adequate funds being forthcoming, subscriptions being freely promised by manufacturing firms and others.

Consideration and discussion soon opened up the beginning of difficulties. The chief of these seemed to be, that sending anything in the way of a present would be quite misunderstood by the Chinese Government and people. The Chinese theory is that China is the middle or central kingdom, and that all nations pay (or ought to pay) tribute to it: to combat this assumption has been one of the chief difficulties of European Ministers in their endeavours to insist on the equality of independent Sovereigns. Any present, and particularly one of great value or extensive appearance, would have been accepted and paraded as tribute. This would clearly have been a great diplomatic sacrifice. Possibly some may be sufficiently enthusiastic to think that it would have been worth this risk. But when responsible Ministers abroad have, after years of effort, succeeded in breaking down a fallacy involving such serious international considerations, it would be too much to expect that any steps should be taken in a contrary direction.

The author's chief idea, in endeavouring to introduce railways into China, was to increase commerce between that country and our own; and he therefore drew attention to the great progress made in our own country during the reign of Her present Majesty, who succeeded to the throne at about the same age as the Emperor of China now was, and that it was surely worth while to make the effort, that the advantages, which had accrued to this country during the reign of Her present Majesty, might be experienced by the Chinese during the reign of their then promising Sovereign.

It is much to be regretted that the scheme was impracticable; for if it could have been carried out, it would most likely have had the effect of increasing the desire, which the Emperor had already manifested, to see more of his people and his country. Condemned to the secluded life of a palace and a garden, and the influence of four or five wives, it is not surprising that he soon found refuge from the troubles of life by quitting it. He died of a violent attack of small-pox, about a year and a half after his assumption of personal government.

N 2

This sad event has thrown the country again into the trials of a long regency. In China the Emperor is succeeded by his son; or, if he has no son, he must be succeeded by some member of the royal family of a younger generation than his own. The present Emperor is five years old; and thus, at the soonest, it must be many years before any personal influence can be wielded by the head of the State.

This particular way of introducing the railway into China had therefore to be given up. Still the author felt that some opportunity would probably arrive for securing the long wished-for step. With a view of being ready for any such chance, he designed and built at the Waterside Works, Ipswich, an engine which should be strong enough to take an appreciable load, able to run 15 or 20 miles an hour, and yet be so small that it could be packed up in a case, and sent out whole to some friend for trial and exhibition in China. This engine was begun in the autumn of 1873, but as there was no hurry about it, it was not finished until the autumn of 1874. At its first trial it did even more than was required, although its weight in working order was only 22 cwt. Mr. John Dixon, who had always taken a warm interest in the project, provided about half a mile of light rails and fastenings to be sent to China with the engine.

Just at this time a proposal for laying a railway on the proposed road from Shanghai to Woosung had been discussed, and given up, for the reason that the estimates for the line amounted to about £100,000; whereas the available capital of the Company, after paying for the land, had become reduced to about £20,000. These estimates were prepared on the spot, by Mr. F. N. Sheppard, on behalf of Mr. Gabrielli, who proposed to construct the line on the full English model. Funds, however, were not sufficient, and the various risks attendant on the project seemed to make the further expenditure of so large a sum as £100,000 unadvisable.

In the spring of 1875, Mr. Macandrew and Mr. F. B. Johnson, two of the directors of the Woosung Road Company, and Mr. Dixon, visited Ipswich to see the little engine. Running on a circle of only 1 chain radius, it maintained a speed of 15 miles an hour. It was also tried with several small trollies laden with iron. Altogether its performance was such, as to establish the opinion of all then present that it would be exactly the kind of engine with which to break the ice in China; large enough to work well—too small to be objected to. A further opportunity of trial was afforded by about a mile of tramway, laid by Colonel Tomline for private purposes, at Felixstowe. Thither the engine was despatched, about Easter 1875, together with three or four trucks. It carried passengers for several days, and during its trials it took frequently forty, and on one occasion eighty, passengers at one trip.

Up to this time the engine had cast-iron wheels, of 18 inches diameter and 2 feet gauge, and cylinders 4″ diameter. As its boiler provided an ample supply of steam, it was deemed advisable to alter the cylinders to 5″ diameter; and as it was now going to China, under circumstances which might involve hard work, it also seemed well to fit it with wrought-iron wheels and steel tyres. These alterations afforded the opportunity of widening the gauge to 2 feet 6 inches. It was also furnished with a larger water tank, though at some sacrifice of its appearance; but this was necessary to enable it to carry water for a trip of a few miles. These alterations brought the weight up to about 30 cwt. in running order; and in view of its now destined service, it was appropriately named the "Pioneer."

The sending out of this little "Pioneer" to Messrs. Jardine's road having been decided on, the author at once endeavoured to see how far the remaining capital could be made to go towards providing a practicable passenger line. For a populous country like China, everyone would have preferred the full English gauge, but everyone equally felt that full gauge and full size were for the time impracticable.

A gauge of 2 feet 6 inches was decided on, partly because the existing engine could be readily altered to that extent, and partly because so fixing it would be certain to secure the opportunity of reconsidering the gauge question, whenever any extension of the line might be in view.

Estimates were prepared by the author to suit the money in hand. Ballast being a very expensive item, it was proposed to have a very liberal supply of sleepers, 2500 to the mile; rails, 20 lbs. per yard; two engines of 6 tons weight each; one first-class, one second, and four third-class carriages; and a supply of plant of all kinds as moderate as possible. Eventually the rails were made 26 lbs. per yard, and the engines stretched to 9 tons.

Even these moderate estimates exceeded the funds in hand. Mr. John Dixon then solved the difficulty, by liberally offering to take a contract to make and equip the line for £20,000 cash and £8000 in shares.

As the undertaking was of such limited dimensions, Mr. Gabriel James Morrison,* M. Inst. C.E., was appointed to the two-fold office of company's engineer and contractor's agent, and Mr. G. B. Bruce kindly consented to act as honorary engineer in England.

The contract was arranged in August, and Mr. Morrison left England on the 1st October, *viâ* New York and San Francisco. At the end of October his five assistants sailed from London for Shanghai direct, in the steamer "Glenroy."

In setting out on such an expedition, these men were quite aware of the personal risks they might have to encounter; on the one hand from probable interference of the authorities, and on the other from possible misunderstandings with large masses of people; to say nothing of the dangers of the climate to persons working long hours, under circumstances involving exposure to weather of all kinds.

Their names were :—

> JOHN SADLER, foreman.
> WILLIAM GEORGE JACKSON, chief working engineer.
> DAVID BANKS, second working engineer.
> JOHN SADLER, jun., second foreman.
> GEORGE SADLER, general assistant.

---

* It is a noteworthy coincidence that the engineer of the first railway in China and the great translator of the English Bible into the Chinese language were both named "Morrison."

Subsequent events showed what the risks were which Mr. Morrison and his staff had to pass through. Mr. Sadler was several times ill of dysentery, and in August had to be sent to Chefoo for change of air. Sad to relate, he never recovered ; he died at Chefoo, September 15th, 1876. This was the melancholy feature of the adventure.

David Banks was tried for the manslaughter of the soldier who committed suicide. This trial was perhaps not serious ; but instead of being tried, he might have been murdered on the spot.

William Jackson twice had his train stopped, in a menacing manner, by a mob of five hundred people.

All suffered much from dysentery, the common malady, arising from the malaria prevalent in the early morning. It has been said that peace has its victories ; these victories involve risks.

The staff sailed in the steamer "Glenroy," the same vessel having on board the little "Pioneer," and about half of the permanent-way materials. This good ship arrived at Shanghai on the 20th December, and Mr. Morrison arrived on the 8th of January.

The rails and sleepers were delivered into river craft, and conveyed to the different parts of the line.

It has already been remarked that the major part of the capital of the Company had been expended in purchasing the land. Of course, in the absence of anything analogous to our Parliamentary powers, this was a matter of personal bargain with each owner ; and when it is mentioned that there were four hundred different owners on the 9 miles, some idea can be formed of the difficulty. Nor was this all, for in many instances ownership consisted chiefly in a grave of some relative more or less valued in life, but highly valued in view of an approaching railway.

John Chinaman is, however, always open to the persuasion of the dollar, and these difficulties were all got over sooner or later by greater or less expenditure. This was managed without any dispute or disturbance. An unfortunate man was summoned before a magistrate and beaten to death, but this was for attempting to defraud a relative out of her share of purchase-money, which had already been paid by the Company ; and this untoward circumstance, although much to be regretted, could not in any way be laid to the charge of the Company. Part of the offence consisted in having sold to the Company a piece of land on the further side of a wide creek at the Woosung end of the line, and thus giving the railway a footing on the further side of what might have been considered as a natural barrier.

Besides purchasing the land, it had also been necessary to proceed with the earthwork to prevent the old owners re-entering and growing crops upon it again, as they would have done if it had not been visibly occupied ! Thus a practicable embankment, about 8 feet

ARRIVAL OF STAFF AND MATERIALS AT WOOSUNG.

Arrival of the first Locomotive in China.

high, had been completed in order to secure possession of the ground bought and paid for.

On arrival of the staff, attention was at once given to the building of about fifteen small wooden bridges over the various creeks. So numerous are the watercourses that, besides the fifteen bridges, upwards of twenty substantial wooden culverts had already been constructed. On the 20th January rail-laying was fairly begun by Mrs. Morrison driving the first spike into the first rail : platelaying and ballasting now progressed rapidly.

The Chinese have always been very skilful in carrying heavy loads by manual labour ; indeed it would almost seem as if their redundant population were the natural substitute for the more scientific appliances of other less populous nations. However heavy the piece to be carried, its transport is only a matter of skilful arrangement of bamboos and ropes, and plenty of help. Moreover, the weights carried per man would be no discredit to the most stalwart English labourer.

The " Pioneer," weighing 26 cwt. empty, was carried by sixteen men, as depicted in the photograph, a distance of three furlongs without stopping to draw breath. This method of reception of the first locomotive arriving in China was especially interesting, as being a practical contradiction to all the assertions which had been made as to the superstitions of the Chinese people on the subject, and their certain opposition. Here sixteen willing labourers carried the first offender, with a zeal and vigour which could not have been surpassed by a like number of the most ardent locomotive superintendents in England.

On the 14th February, 1876, the " Pioneer " made its first trip on about three-quarters of a mile of rails. The news was telegraphed to England, and was received in London the same day :—" Engine ran to-day. Chinese delighted." A very pleasant valentine for the promoters !

The delight of the Chinese people was unbounded ; numbers of people flocked to see the little engine at work, and the interest manifested in the railway was continually on the increase. This was viewed with great alarm by the Taotai of Shanghai ; and about the 23rd of February, he was so pressing in his demands for a discontinuance of the work, that a compromise had to be made ; to the effect that the running of the engine should be discontinued for a month, but the works were to proceed until he should receive definite instructions from Pekin.

The month expired, and no further interference occurring, it was presumed that at any rate no adverse instructions had come from Pekin, and the engine resumed work in the latter part of March. The interest, which was great before, was very much quickened by the stoppage and the resumption of the " Pioneer's " trips. Much in the same way as when the Lord Chamberlain complains of a play half London goes to see it, so the Chinese, having been told to have nothing to do with the railway, now flocked in thousands to the spot.

The following letter from ' The Times' correspondent at Shanghai, under date March 31, 1876, appeared in that journal on the 22nd of May, and gives the experience of an eye-witness :—

" You will be glad to hear that the construction of the little Woosung railway is progressing, and there are symptoms of withdrawal of opposition on the part of the Chinese officials. It is rumoured that a hint was received by them a few days ago from Pekin to see as little as they could of what was happening, and straws seem to confirm this hint of a change of wind.

" The persecution to which I have before referred of people who had sold certain pieces of land has ceased, and one or two plots which the Mandarins show some reason for wishing to recover are likely to be amicably exchanged ; for instance, one which touches the river embankment will be readily exchanged for an adjacent piece a little inland, and the piece on the opposite side of the Woosung Creek, to which I referred a few weeks ago as a cause of trouble, will also be surrendered.

" In the meantime there is no interference with the workmen, who are all country people, and things are progressing rapidly. Several miles of road have been completed and ballasted, and the whole country side is alive with interest. Literally thousands of people from all the neighbouring towns and villages crowd down every day to watch proceedings, and criticise every item, from the little engine down to the pebbles of the ballast. All are perfectly good humoured and evidently intent on a pleasant day's outing. Old men and children, old women and maidens, literati, artisans and peasants—every class of society is represented ; and enterprising peep-show and fruit-stall men have taken advantage of the opportunity to establish a small fair on a convenient spot in the neighbourhood.

" The engine, of course, is the great centre of attraction. It is engaged in dragging trucks with pebble ballast at present, and a general cry of " Laij tze, laij tze !"—"It's coming, it's coming !" heralds each return journey. Then ensues a crowding around, and an amount of introspection which suggests awful reflections in case of accident, and then the whistled signal to start ; the fall of a live shell could hardly suggest a greater stampede, except that laughter and perfect good temper are present instead of terror.

" Everything, therefore, is going on so far satisfactorily ; and if the people are let alone by their officials, they will quietly satisfy their curiosity and go home amused and interested. They are giving practical proof at present of what I have always urged—that there is no instinctive dislike in the masses to things foreign. There is only a great deal of ignorance, which can easily be played upon by the officials, and dangerously misdirected if it suits their purpose. Let us hope that this little pioneer railway will get finished without further trouble, and that it will serve to introduce into China a mode of carriage which has done so much to develope the resources of western countries."

About the same time the following notice was inserted in the Shanghai newspapers :—

" In consequence of the crowds of people who assemble daily to stare at the progress of the Tramway, the Municipal Council very wisely desire that rifle practice at the Butts shall cease for the present. Any accident would not only be regretable in itself, but in the last degree unfortunate from a political point of view, under the circumstances. If no misfortune happens, and the people are not interfered with, they will gaze their fill, and go home with their curiosity satisfied and a clear idea that railways *are not very awful things after all.*"

At every return trip of the empty ballast trucks they were re-crowded with gratuitous passengers. Nor was this desire to try the railway limited to the lower orders. Several times during the progress of the works, Chinese personages of high rank came, arrayed in their best clothes, to take a ride in the ballast trucks. No other accommodation was available, but on such occasions Mr. Morrison would have seats placed in the trucks covered with red baize, and a carpet laid in the truck of the same material.

As the works approached completion, various sinister rumours began to float in the air as to the intentions of the governing powers, and a very politic step was taken in inviting various Chinese notabilities and all the foreign consuls to take an excursion trip; they all attended in full uniform to give the full weight of their official position in the right direction. This event was recorded by the 'North China Herald' as follows :—

"The first railway excursion train in China was run on the 26th May, 1876; the excursionists including several ladies, who were accompanied by Admiral Lambert and a party of gentlemen. The train was composed of five ballast trucks, carpeted and otherwise properly furnished for the occasion. The distance traversed extended over about five miles, and the trip was thoroughly enjoyed."

All continued to go well, and the works were not interfered with. The Company's engineers were continually besieged by applicants for the post of driver, the argument generally being that as the candidate had assisted in driving a steamboat on the river, he could equally well drive " the steamboat on shore," as the embryo train was called. At a later stage this title was changed to the "fire dragon carriages :" but this was after the permanent engines and carriages had arrived.

These permanent engines, though very small according to our ideas (being only 9 tons in working order), were large indeed compared with the "Pioneer," and were esteemed by the Chinese accordingly. On the 30th May the first permanent engine, the "Celestial Empire," arrived, and in a few days it was put together, and it made its first trip on the 12th June to Kangwan. This trip it performed at a speed of 25 miles an hour—rather a high speed for a six-wheeled coupled engine, with wheels of only 27 inches diameter.

The carriages arrived about the same time. They were about half the length, two-thirds of the width, and three-fourths of the height of English railway carriages. In the first instance the rolling stock consisted of

1 first-class carriage, 15 ft. long, accommodating	16 passengers.
1 second-class ,, 15 ft. ,, ,, ...	18 ,,
4 third-class ,, 18 ft. ,, ,, ...	96 ,,
Total	130 ,,

and 12 trucks measuring 10 feet by 5 feet by 1 foot 6 inches, weighing empty 25 cwt., and equal to a gross load of 5 tons.

Three classes of carriage had been determined on, because the first-class passengers would be willing to pay very high fares, and it was desirable that the third-class should only pay very low ones. It therefore seemed necessary to have an intermediate class.

In practice, however, it was found that many trains required third-class carriages only; the proportion of passengers of different classes was one first and two second-class passengers to eighty third-class.

O

The success achieved in Japan in conducting the railway business in the English language, led the directors to decide on the same experiment in China. The railway tickets were therefore printed in English, but Chinese characters were printed on the reverse side, indicative of the stations. This proved to be quite satisfactory.

The contract with Mr. Dixon had specified the 1st July as the day for opening the railway. On the 30th June about 150 of the European residents were invited to make the opening trip in the permanent train, over the first five miles of line from Shanghai to Kangwan.

This was successfully performed, in a train laden with about 200 passengers, at the speed of 15 miles an hour. Short as the journey was, refreshments were of course provided at Kangwan, and the Company, the engineer, and the contractor were all duly toasted in the usual English fashion.

On Saturday the 1st July, the Chinese were invited to travel free on the railway— great, indeed, was the jostling, pushing, and crowding to secure places in so small a means of conveyance. All, however, passed off well and without accident, and to the great delight of the people. On the 3rd July traffic commenced, and the receipts were at once of the most satisfactory character.

Just at this time the telegraph to England was unfortunately interrupted, and so the news was not received in England until the 6th July. This first intelligence of a real railway (though a small one), open for public traffic in the hitherto sealed empire of China, was naturally the cause of great rejoicing on the part of the directors and promoters of the undertaking, and a dinner (without which no English undertaking can be considered complete) was held at the Langham Hotel on the 20th July to celebrate the event.

The day was auspicious, for the telegraph brought in the course of the afternoon a demand from China for another complete set of carriages. This request, coming on the very day of the festival, was extremely opportune, and seemed to be an earnest of fulfilment of all the best hopes of the subscribers. Of course the carriages were soon made and sent out.

Subsequent mails brought most satisfactory accounts of the traffic. On festivals and holydays it was quite common for the train (of 130 seats) to start with 250 passengers and leave more than that number behind. Extension of the line to Soochow was freely advocated by the newspapers (both English and Chinese), and was generally spoken of as a step both desirable and probable.

The train service gave six trains per day each way, or as many as could be run with one engine in steam between 7 A.M. and 6 P.M.; and the crowded state of some of the trains can be imagined, when it is stated that the third-class carriages *averaged* a full load.

CHINESE CROWD AT THE OPENING OF THE RAILWAY.

The "Pioneer" and the Permanent Train on the Opening Day.

The following is the first railway time table issued in China :—

WOOSUNG ROAD CO., LIMITED.

ON AND AFTER

MONDAY, JULY 3rd, 1876,

UNTIL FURTHER NOTICE, TRAINS WILL RUN ON THIS LINE

BETWEEN

SHANGHAI & KANGWAN

AS UNDER :—

Trains will leave Shanghai at

A.M.	A.M.	A.M.	P.M.	P.M.	P.M.
7	9	*11	*1	3	5

Trains will leave Kangwan at

A.M.	A.M.	A.M.	P.M.	P.M.	P.M.
7.30	9.30	*11.30	*1.30	3.30	6

FARES :

	SINGLE TICKET.		RETURN TICKET.
FIRST CLASS	$0·50	..	$1·00
SECOND CLASS, 300 Cash, or	0.'25	..	600 Cash, or 0·50
THIRD CLASS, 120 Cash, or 10 for	1·00	..	200 Cash, or 6 for 1·00

The Full Number of Good Cash will be demanded. Children under 10 Years will be charged Half Fare.

Trains marked thus * will not run on Sundays.

O 2

All went well and without accident until August 3rd, when an unfortunate lunatic committed suicide on the line.

That it was a case of suicide was very graphically described at the time by one of the Chinese brakesmen on the train, in the following broken English :—

"My have see one piecey Chinaman. My tink he all same belong. Soldier man, he makee walkee on that lailway. All same time my piecey train come that side. Mr. Ban-kas makee that engine whistle plenty long time. Then he makee go off lailway litty time. My tink he must wantshee makee die, caus he makee come that side number two time. My have see that piecey engine hit he. My have all same time puttee on that piecey brake—makee that train stop ' chop chop ' " (quickly).

So prompt indeed had our friend been with his brake, that the train was pulled up in a very few yards. The suicide had stepped off the line in obedience to the whistle ; and, still walking towards the train, he stepped on to the track again, when the engine was about six yards from him. Of all this there were several witnesses, who were able to place the matter beyond dispute.

The case was, of course, duly reported to the authorities, and the Taotai rejoined that he knew there were five hundred men anxious to commit suicide on the railway—an announcement which might have been very alarming, had it not been perfectly well known how very common the crime of suicide is in China ; and it was judged that the Taotai was rather speaking of the railway as an expeditious and novel method, than of his certain knowledge on the subject. No doubt he wished that all persons who might have such serious intentions, should adopt the railway as their means to their desired end. However this may be, no other cases occurred.

Curious to relate, no one appeared to claim damages for the death of the deceased, nor even could any evidence be obtained as to his name or identity in any way, although he was said to be dressed like a soldier. He appeared to have been without means, without friends, and alone in the world ; and perhaps these were his reasons for desiring to leave it. It was at any rate clear that he had not been hired for the purpose ; for in that case, relatives would have been produced to claim damages, and to assist in the clamour.

It seemed advisable to put the driver on his trial before the Consular Court, as a matter of protection to him. Of course he was honourably acquitted, and costs were given against the Court.

One day a slight disturbance arose at Kangwan from a station master having mis-understood his instructions as to a fence and a right of way claimed by an adjoining occupant ; but this was at once set right by the promptitude and tact of Mr. Morrison.

In the month of July Mr. Wade (now Sir Thomas Wade) was engaged in an endeavour to settle the " Margary " dispute ; and about the middle of that month he left Pekin for Shanghai. Subsequently a meeting of plenipotentiaries was arranged to be held in Chefoo, in August or September. Before negotiations commenced, Mr. Wade found that there was a predisposition in the minds of the Chinese, to consider the murder of Mr. Margary as balanced by the death of the Chinese suicide on the railway. " We have killed one of your men, and you have killed one of ours," was the argument.

This was in itself sufficiently alarming, and naturally suggested to Mr. Wade the fear that should any really serious accident happen, untold complications might arise. He therefore sent a letter, by special steamer from Chefoo, to the Railway Committee at Shanghai, to request that the trains should cease running pending his negotiations.

Under all the circumstances, the only patriotic course was to accede to a request, which it was felt ought to have all the force of a command, and on the 23rd August the railway was consequently closed.

In the month of September Mr. Wade completed the now-celebrated Chefoo Convention; and soon afterwards negotiations were commenced on the subject of the railway.

It was now found what a mistake had been made in closing it. Its first position was simply unassailable on any just grounds. The promoters had bought the land, and surely had a right to lay their rails and run their own trains on their own property !

True, it was urged, that in all other countries railways are under more or less of government control. But in China there were no laws whatever on the subject of railways; and in the absence of prohibition, an act, which is innocent in itself, is always lawful. But the railway had been closed pending an international discussion, and this virtually placed the whole undertaking at the mercy of the Chinese.

The terms eventually made were that the provincial government should buy the railway at a price corresponding with its cost; that the money should be paid in three half-yearly instalments, and that until the whole should be paid the Company should work the line. These arrangements were made with the assistance of Mr. Mayers, Secretary of the British Legation, and he received, deservedly, great credit for arranging the matter fairly to the satisfaction of both sides.

The first instalment was paid down, and this gave the railway just one year of grace for practical work. This being secured, the other terms of the bargain were not so closely scanned;—the desideratum was to give China one year of railway experience, and that being accomplished, it was hoped that only one result could follow. Even the condition that no goods were to be carried was not seriously dwelt upon, although this stipulation was equivalent to depriving the railway of more than half its earning power.

It was hoped that the ownership of the line being vested in the Chinese might have a good effect, and that the authorities might feel more interest in an undertaking which had now become their own. At first this seemed likely to result, and there was some talk of Tong King Sing being appointed general manager of the line after the year of grace should expire; and, as he was already a shareholder in the railway, and general manager of the China Merchants' Steamboat Company, this was deemed a hopeful sign. The Taotai issued the following proclamation on the subject of the railway :—

WOOSUNG ROAD CO., LIMITED.

TRANSLATION.

PROCLAMATION BY FENG TAOTAI AND SUPERINTENDENT OF CUSTOMS, &c.

In the Matter of the Railway between Shanghai and Woosung certain terms have lately been agreed to at Nanking, by Mr. Secretary Mayers with Taotais Chü and Sheng and myself, by which the line is bought up by the Chinese Government, and the purchase money is made payable between the 15th day of the 9th month of the 2nd year of Kwang-hsu (21st October, 1876), and the same day of the ensuing year.

The payments being thus left outstanding, it was considered only fair that during the interval the Company should be left the option of running the trains, and it was arranged that regulations for the safer conduct of the traffic should be framed by the Consul and Taotai together. All this was duly reported to the Commissioners for Foreign Trade (Northern and Southern) and to the Governor of Kiangsu, all of whom have since signified their assent to the terms. In consequence I, the Taotai, have, in consultation with H. B. M.'s Consul agreed upon Six Rules, which have been severally communicated to the District Magistrates and deputed officers concerned, and which it is further necessary to make public for general information.

Wherefore I hereby promulgate the said Rules for the information of all people, civilian and military alike, and call upon them to yield implicit obedience thereto.

RULES.

1. Fences or ditches will skirt the line on both sides. Where public roads intersect it, gates will be erected for the convenience of wayfarers and animals passing to and fro.

2. The gates thus erected having to be opened and closed from time to time, watchmen will be stationed outside thereof to look after them day and night. The trains having to run at specified hours, the watchmen must close the gates before the train starts, and only open them after it passes. Should any watchman happen to be negligent and not close his gate previous to the starting of a train, so as to induce people to attempt to rush through and possibly fail in escaping the passing train, it shall become the duty of the Company to see the matter properly looked into and adjusted.

3. The gates at public road intersections being once closed for an approaching train, should any person insist on pushing through, he may be taken charge of and sent to a deputed officer's quarters, whence he will be forwarded to the Mixed Court at Shanghai in order to be rigorously punished. But the gate-keeper must not take upon himself to beat the man, lest disturbance should ensue.

4. There being old public roads on both sides of the railway line, for the use of travellers, no person will be allowed to trespass upon that portion of the line between the fences

or ditches, save and except for the purpose of crossing the line where gates are provided at public road intersections. Neither may horses nor cattle be led therein to tread it down. Any breach of this Regulation will render the parties liable to be handed over to a deputed officer for punishment.

5. At the Stations appointed for the issue and receipt of tickets, the authorities will place persons for purposes of scrutiny. Passengers will be allowed to carry with them baggage and little sundries, but it will not be permitted to convey goods of any kind to the evasion of duties or imposts. Should any attempt be made to carry concealed opium and such like articles, rigorous punishment and confiscation will be the certain result on discovery. On the other hand the detectives will not be allowed purposely to annoy passengers.

6. Trespass upon the Station premises or on any part of the Railway property will not be permitted. Persons so offending will be handed over to a deputed officer for punishment.

KWANG-HSU, *2nd year, 10th month, 10th day.*
      25th November, 1876.

The Directors beg to call special attention to Rules IV. and VI., as in the interest of public safety it is necessary that they should be strictly observed both by Foreigners and Chinese.

The following rules and regulations were also published as bye-laws of the Company :—

WOOSUNG ROAD CO., LIMITED.

SHANGHAI AND WOOSUNG RAILWAY.

I. No person shall be allowed to travel by any train until he has obtained a ticket.

II. No person shall be allowed to travel beyond the Station for which he has a ticket.

III. No person shall be allowed to travel in a carriage of a superior class to that for which he has a ticket.

IV. Any person found travelling without a ticket will be charged the full fare from the point where the train started, with 10 per cent. additional ; and any person found travelling beyond the Station for which he has obtained a ticket, or in a superior Class to that for which he holds a ticket, will be charged the difference of fare, with 10 per cent. additional.

V. None but Passengers shall be admitted to any of the Stations.

VI. No person shall be permitted to ride on the platform of any carriage, or to enter or leave a train while it is in motion.

VII. Each passenger shall be allowed half a picul of personal luggage free, but no passenger shall be allowed to infringe Rule V. of the Taotai's regulations or to take any article into a carriage which is likely to cause inconvenience to the other passengers.

GENERAL MANAGER'S OFFICE.
    SHANGHAI, *December 1st,* 1876.

The additional charges of 10 per cent. levied on persons neglecting to take tickets were made with the view of making the people careful about such matters from the outset. As the Chinese people always go the cheapest way to work, this regulation was quite successful. The limitation as to the character of personal belongings allowed to be taken into the carriages was directed against the market gardeners; after selling their produce in the towns, they return to the country with their buckets filled with manure.

The new time table, adopted at the re-opening, was more extended than its predecessor :—

WOOSUNG ROAD CO., LIMITED.

SHANGHAI & WOOSUNG RAILWAY TIME-TABLE.

TRAINS WILL RUN AS FOLLOWS:—

	A.M.		A.M.		A.M.†		P.M.†		P.M.†		P.M.		P.M.	
Leave Shanghai	6	..	8	..	10	..	12	..	2	..	4	..	6	..
,, Kangwan	6	17	8	17	10	17	12	17	2	17	4	17	6	17
,, Woosung Bar	6	35	8	35	10	35	12	35	2	35	4	35	6	35
Arrive at Woosung Creek ..	6	40	8	40	10	40	12	40	2	40	4	40	6	40
Leave Woosung Creek	7	..	9	..	11	..	1	..	3	..	5	..	7	..
,, Woosung Bar	7	5	9	5	11	5	1	5	3	5	5	5	7	5
,, Kangwan	7	20	9	20	11	20	1	20	3	20	5	20	7	20
Arrive at Shanghai	7	35	9	35	11	35	1	35	3	35	5	35	7	35

FARES FROM SHANGHAI.

SINGLE TICKET.				RETURN TICKET.		
1st Class.	2nd Class.	3rd Class.		1st Class.	2nd Class.	3rd Class.
$ c.	$ c.	*cash*		$ c.	$ c.	*cash*
0·50	0·25	100	To Kangwan	0·75	0·50	180
1·00	0·50	200	,, Woosung Bar	1·50	0·75	360
1·00	0·50	200	,, Woosung Creek	1·50	0·75	360

FARES FROM WOOSUNG.

SINGLE TICKET.				RETURN TICKET.		
1st Class.	2nd Class.	3rd Class.		1st Class.	2nd Class.	3rd Class.
$ c.	$ c.	*cash*		$ c.	$ c.	*cash*
0·10	0·05	30	To Woosung Bar	0·20	0·10	50
0·50	0·25	100	,, Kangwan	0·75	0·50	180
1·00	0·50	200	,, Shanghai	1·50	0·75	360

CHILDREN UNDER 10 YEARS WILL BE CHARGED HALF FARE. DOGS CHARGED 10 CENTS EACH FOR ANY DISTANCE.

The full number of good Cash will be demanded. Exchange 1200 Cash per Dollar.

Trains marked thus † will not run on Sundays.

RAILWAY TRAIN FILLED WITH EUROPEAN AND AMERICAN VISITORS.

RAILWAY STATION AT WOOSUNG CREEK, WITH ORDINARY TRAIN.

The copper cash in use by the Chinese are so minute, that for some tickets 1200 had to be counted ; this involved so much labour, that frequently four or five booking clerks were required in the Shanghai office to get through the counting of the money. To obviate this to some extent tickets were issued so many to the dollar, according to value.

The second opening of the line took place on the 1st of December—this time the whole distance to Woosung. On this occasion many Chinese of high rank attended, and drank success to the undertaking with a hearty good will in champagne provided by the Company. The Taotai and his officers were of course invited, and for some time it appeared as if they intended to come, but they finally decided not to take such a very serious step.

The supply of engines and carriages was so limited, that only a single train could be provided as à rule, and as the journey occupied a little more than half an hour, the train could only make seven double trips in the day. The interval between one train and the next being two hours, was, of course, a great disadvantage in what was essentially suburban traffic, although time is not of so much importance with the Chinese as it is in Europe.

Shortly after the re-opening of the line, the driver of the first train in the morning was rather alarmed on arriving at Woosung to find a great crowd of people standing on the line in a threatening manner. He very promptly decided that there were too many of them to be intending suicide, so he boldly put on extra steam, and the crowd at once opened out. This disclosed a heap of ballast piled on the line ; fortunately it was not high, and the train crushed through it without accident.

Arrived at Woosung it was found that the reason of the disturbance was, that the people at Woosung did not approve of platelayers from Kangwan being employed at Woosung. During the construction of the earthworks the people had been so anxious for employment on the railway, that a sort of arrangement had been made that the people of each village should be employed to make the portion of earthwork in their own district. After the completion of the line, however, this arrangement was impracticable, as only a few skilled men would now be required to keep the line in order. As the train was leaving Woosung on its return journey, the conductor very cleverly captured two or three of the ringleaders, and locked them in a carriage just as the train was moving off; he took them to Shanghai and lodged them in prison, and this had a salutary effect on the other rioters. The matter was soon fully explained, and no further disturbance occurred.

Not only were the people anxious for employment on the railway, but they manifested very great aptitude. The platelayers soon became quite equal to English workmen ; and as breaksmen and firemen the Chinese were all that could be desired. Some of them were also very passable as fitters and turners, and in course of time there is no doubt they would make excellent drivers.

In this connection it may also be mentioned that more than half the shareholders (in number) were Chinese. In short, nothing like hostility was ever manifested by the people. Such little difficulties as there were arose from misunderstandings, and were always at once adjusted by simple explanation.

P

The trains continued to run steadily and well, the traffic fluctuating a good deal with the weather, but it frequently amounted to £27 per mile per week, thus nearly equalling the average receipts from passenger traffic on English railways, although there was only one train at work.

The shortness of the line, of course, caused the expenses per mile per week to be very heavy, as the highly-paid European staff would have sufficed equally well for a line four or five times the length, and the native staff of gatemen, watchmen, &c., &c., was of necessity very numerous in the particular locality close to Shanghai.

Still, notwithstanding all the disadvantages of crippled conditions as to receipts and severe conditions of expenditure, the line was worked at a profit. The behaviour of the people was uniformly good, everyone regarding the railway in the most good-humoured and good-natured way. The limited tenure of the Company, of course, rendered it unadvisable to incur great expense in furnishing additional engines and carriages, and the increase of stock was limited to one locomotive, and the six carriages ordered in the summer.

At first it was hoped that the working of the line would be continued by the Chinese after the completion of its purchase, but as month after month passed away without any definite arrangements being made, this hope gradually disappeared. Mr. Morrison made special visits to His Excellency Li Hung Chang, and other persons of influence and power, but without success. The consuls and governments of various nations made representations on the subject, one of them going so far as to say that the stopping of the railway would be regarded as a " most unfriendly act."

By this time too the Chinese ministers Kuo Ta-jen and Liew Ta-jen had arrived in England, and they also sent the most urgent solicitations, both to their imperial government and to the Viceroy of Nankin, to refrain from such a retrogressive step.*

In the spring of 1877 Mr. Morrison paid a visit to the island of Formosa, at the request of the Governor Ting Futai, who expressed his desire to make a railway from one end of the island to the other.

The views of Ting Futai are of the most enlightened type, and if he were viceroy of a province on the main land there would be no difficulty in carrying out his wishes. But to make railways in the island of Formosa would not be easy, as there are both physical and political obstacles.

---

* One of the first journeys taken by their Excellencies in England was in the company of the author, and their intelligent appreciation of everything connected with the railway was a pleasing contrast to the attitude of the authorities in China. That they should be impressed with the ease, the comfort, and the speed of the railway was perhaps only what was to be expected, but it was one more illustration that the railway needs only to be practically tried to be approved. The author will never forget the forcible eloquence with which Liew Ta-jen reviewed the work of the day and its two long railway journeys, declaring that a similar expedition in his own country would have taken six days instead of one. It made one wish that the Viceroy of Nankin were of the party !

Engine and Carriage Sheds at Shanghai.

All things considered, even a successful railway in Formosa would not materially advance the cause on the continent of China, whilst an unsuccessful one would certainly retard it. Two hundred miles of railway might be made a success in Formosa in a few years, but this would be a mere fraction of what may be expected from perseverance on the main land. For instance, the influence which Europeans have brought to bear on the Chinese from Shanghai, as a mainland settlement, has far exceeded that of Hong-Kong. The circumstance of Hong-Kong being an island no doubt facilitated our first acquisition of it as a British settlement, but the intervening sea has always proved a serious barrier.

In the same way Formosa would not afford the favourable field for introducing railways into China, which a similar opportunity on the continent of China proper would do. The author has from the beginning, in common with others, advocated special efforts with the distinct object of bringing China into closer relationship with the rest of the world. The population of that country is one-third part of the whole world, but its commerce is infinitesimal. The whole imports and exports of China amount to about 2s. 6d. per head of her population, whilst the imports and exports of Great Britain amount to £12 or £14 per head of population.

Again, the commerce of Great Britain with China is at the rate of about 2s. per head of the Chinese population, whereas our commerce with all the rest of the world is at the rate of 12s. per head. Thus our commerce with China is only one-sixth part of the average. The one thing which is required to redress this discrepancy is a system of railways in China. Each Chinaman at present sells to this country about 1s. 3d. worth of his goods per annum, and buys from us about 9d. worth of ours. Railways would certainly give our manufactures the *entrée* in such quantities, that they would rapidly penetrate to districts which they now never reach.

Railways would on the one hand relieve our manufactures from the vexatious imposition and reimposition of local taxes, and on the other hand would afford to the government the most certain means of collecting just dues. Everyone conversant with railway working knows that accurate account-keeping is a first necessity; the quantity and nature of goods passing must be always ascertained, and recorded in such a way that every transaction can at any time be fully ascertained, even years after its occurrence.

Under a mistaken idea on this subject, the Chinese authorities refused to allow goods to be carried on the Woosung line. It is interesting to note that at this very time another government, in a similar situation, is making a railway for the express purpose of obtaining control over all imports, and is insisting on all goods being sent into its territory by railway, and by no other means of conveyance.

The desire of Ting Futai to possess a railway has, of course, been perfectly well known in China for a long time, and this has led to the proposal to remove the present railway to Formosa. Fancy removing a railway! How strange it sounds! The reasoning is, however, very direct. One man has a railway which he does not want, and another wants a railway but has none; so let the man who does not want the railway send it to the man who wishes for it!

During the spring and summer of 1877 the trains continued running without any interruption, and, fortunately, without any accident. This was matter of great congratulation, as the stations and trains were often crowded to excess, and casualties might easily have happened.

Only once was anything set on fire by sparks from the engine, and this was in October 1877. The damage done was trifling, as may be judged from the fact that a sum of £90 was accepted in full of all demands. Some rice straw was set on fire, and this extended to a house close to the line. Although plenty of help was immediately given, the house was of such a flimsy character that it was burnt down in a few minutes.

On the approach of the next train to the spot (about a mile from Shanghai), the driver found about a thousand people congregated on the line, in much the same way as they had assembled on the occasion before described at Woosung. He, of course, at once assured the people that Mr. Morrison would come by the next train, and pay them some dollars.

Although thus ready in lending their aid to "squeeze" the Company, none of these people were willing to give shelter to the occupants of the house, and they had to pass the first night in the fields. It appears that the Chinese have a superstition that persons burnt out of their house carry ill-luck about with them until after rain shall have fallen. This coming to Mr. Morrison's ears the next day, he provided the luckless folk with lodgings and refreshment at Shanghai. On the third day rain fell, and they were then able to get house-room nearer their former home without difficulty. As usual in such cases, the compensation not only covered the damage done, but all sorts of collateral claims as well.

The time for giving up the railway now drew on rapidly. Everyone along the line was interested in the continuance of the train service.

Instead of the railway disturbing the spirits of the earth or of the air, as had been suggested, it was found that the value of property had increased.

Instead of the boatmen and wheelbarrows being thrown out of work by the railway, their employment was doubled.

The people from the neighbouring villages resorted to Kangwan as a convenient station at which to take the train for Shanghai, and they were thus enabled to get better prices for their produce.

From Woosung all classes of people availed themselves of the railway as their ordinary means of transit; the regularity of the service, in all weathers, being considered by them of even more importance than either its cheapness or its speed.

There is in China at all times great unwillingness to take any step which might be interpreted as an act of hostility to any constituted authority. Notwithstanding this, a numerously-signed memorial was forwarded to the Viceroy, as follows :—

(*Translation.*)

THE PETITION OF THE MERCHANTS AND PEOPLE OF SHANGHAI, KANGWAN, AND WOOSUNG.

We, the undersigned, beg to crave that your Excellency would condescend to consider our combined request and representation on this matter.

Since the trains commenced running on the Shanghai and Woosung Railway on the 6th moon of last year, the travelling public has gladly availed itself of the facilities afforded for rapid transit to and fro.

Your Excellency's predecessor having addressed their Excellencies the Governor and Governor-General, they, after communicating with the Tsung-li-Yamen, purchased the line and fixed upon the 15th day of the 9th moon (21st October, 1877) to take over the road, track, carriages, and all fixtures.

The time fixed is approaching, and we do not know whether, after delivery is made, the trains will continue running as usual, or whether they will be stopped and the line removed. As the line has been purchased, there appears to be no harm in allowing it to be worked, so that the public may have the advantage of travelling, and sending their goods. Its being taken up would be a useless throwing away of a large sum of money.

For this reason we jointly beg to place our views before you, and to crave your Excellency's favourable and condescending consideration, and to beg you to represent the matter to their Excellencies the Governor and Governor-General, asking them to depute able officers to manage the working of the line properly, so that the real benefit may be ensured. It is actually a public convenience.

We await your Excellency's reply to our humble petition.

Signed by 145 persons or hongs.

To this memorial no reply was received.

On the 20th October the final instalment of the purchase money was paid, and the line passed into the hands of the Chinese authorities, much to the regret of the Chinese people.

On the 19th October Mandarins, commissioned for the purpose, intimated to the officers of the Company that they would be ready to pay over the balance of the purchase-money the next day. It was then arranged that the last train should run from Shanghai at noon, and from Woosung at one o'clock, on the 20th. This train was the longest and heaviest laden which had traversed the railway, and the sides of the line were thronged with spectators. At two o'clock the last instalment was paid, the formal handing over of the line being postponed until Monday the 22nd.

Of course a special train was in readiness for the Mandarins, and equally of course, was declined by them. They preferred to go in their chairs. Their journey occupied three hours in going to Woosung, and nobody knows how many hours in returning to Shanghai.

This preference for their own "more dignified" means of travel was perfectly consistent on their part. Happily they were quite unconscious of the illustration they were affording of the contrast of the views of the governed and the governing. The closing scene showed the people crowding the railway, and their rulers preferring sedan chairs!

This is probably the first railway which has been completed, worked at a profit for twelve months, and then been bought and paid for in hard cash, for the express purpose of stopping it.

For the present our friends at Kangwan and Woosung must return to their former mode of conveyance. This, however, surely cannot be for long. The line may even be taken up, and removed to Formosa, but the seed has been sown, and it will bear its fruit.

At the moment of going to press the following paragraph from the *Sinpaou*, a Chinese newspaper, usually reputed to be the Taotai's organ, has been received by the mail :—

" The railway being done away with, passengers to and from Woosung have reverted to the passage boats. On Oct. 25 last, at 3 P.M., when two of these boats had arrived off Kao-chiaou, a sudden squall sprang up, and they one after the other capsized, the hundred and odd passengers floating down with the tide shrieking for help. Word was at once sent to the Humane Society, whose boats speeded to the spot. How many were rescued, and how many joined the drowned men's ghosts down below, is not yet accurately known."

Sometimes the most drastic medicine has the best effect ; certain it is that the closing of this line will have caused it to be much more talked about and thought about in China, and the time is probably not far distant when some further progress may be made of a very decided character.

ENGRAVED AND PRINTED BY WILLIAM J. WELCH, 25, WELLINGTON STREET, STRAND, LONDON, W.C.

CHINESE TRAVELLING.

SCALE FOR ESTIMATES, 1, 2, AND 3.

Tons for Freight.				@	
Measure-ment.	Dead Weight.	FROM _____ TO _____		£ s. d.	£

ESTIMATE.

For _____ Miles.   Gauge _____

			£ s. d.	£
R per yard, with fish plates	Tons	..	..	..
F spikes, and fang bolts	,,	..	..	..
S × in. × ft. in. (per mile) ..	No.	..	..	..
P ft. tongues ft. stocks	Sets	..	..	..
C with check rails	,,	..	..	..
P ls and trollies	,,	..	..	..
T turning tank engines (centre balance) ft.	No.	..	..	..
P ing-machines	,,	..	..	..
W tons), marked in English	,,	..	..	..
H with two brass barrels, fly-wheel, pipes, &c.	,,	..	..	..
T ron plates, with planed ⎫ to hold galls.	,,			
rted on columns, with ⎬ ,, ,,	,,	..	..	..
leather hose, &c. .. ⎭ ,, ,,	,,			
W ngs, engine, boiler, shafting, lathes and ces, fan and smiths' fires, taps and dies, drills, &c.	,,	..	..	..
S oil and tallow boxes, rope, chain, spun-ks, screw jacks, &c., &c.	,,	..	..	..
E er fire-box, brass tubes, steel tyres, with ols complete	,,	..	..	..
E tes	Sets	..	..	..
G ns, fitted for luggage and parcels	No.	..	..	..
O gons, ft. in. × ft. in. × ft. in. ; y, cwt. ; full, tons	,,	..	..	..
O gons, ft. in. × ft. in. × ft. in. ; y, cwt. ; full, tons	,,	..	..	..
E wheels and axles, axle boxes, springs, per set	,,	..	..	..
N ders for ft. span .. .. per pair		..	..	..
,, ...		..	..	..
ists ,, ...		..	..	..

T ght.

T nt.

_____

TAL COST of all materials and fittings for miles (including sleepers)   £ _____

,, ,, per mile .. .. £

ULE FOR ESTIMATES, 1, 2, AND 3.

Tons for Freight.						
Measurement.	Dead Weight.	FROM_____TO_____			@ £ s. d.	£

ESTIMATE.

For_____Miles.   Gauge_____

			£ s. d.	£
per yard, with fish plates	Tons	..	..	..
spikes, and fang bolts	,,	..	..	..
× in. × ft. in. (per mile) ..	No.	..	..	..
ft. tongues ft. stocks	Sets	..	..	..
with check rails	,,	..	..	..
ols and trollies	,,	..	..	..
turning tank engines (centre balance) ft.	No.	..	..	..
ing-machines	,,	..	..	..
tons), marked in English	,,	..	..	..
with two brass barrels, fly-wheel, pipes, &c.	,,	..	..	..
ron plates, with planed) to hold galls.	,,			
rted on columns, with } ,, ,,	,,	..	..	..
, leather hose, &c. ..) ,, ,,	,,			
lings, engine, boiler, shafting, lathes and ices, fan and smiths' fires, taps and dies, drills, &c.	,,	..	..	..
oil and tallow boxes, rope, chain, spun- ks, screw jacks, &c., &c.	,,	..	..	..
er fire-box, brass tubes, steel tyres, with ols complete	,,	..	..	..
tes	Sets	..	..	..
ans, fitted for luggage and parcels	No.	..	..	..
gons, ft. in. × ft. in. × ft. in. ; ty, cwt. ; full, tons	6	..	..	..
gons, ft. in. × ft. in. × ft. in. ; ty, cwt. ; full, tons	,,	..	..	..
wheels and axles, axle boxes, springs, per set	,,	..	..	..
ders for ft. span .. .. per pair		..	..	..
,, ...		..	..	..
ists ,, ...		..	..	..

ght.

nt.

TAL COST of all materials and fittings for   miles (including sleepers)   £_____

,,   ,,   per mile .. .. £

Tons for Freight.						
Measure-ment.	Dead Weight.					

ᴿOM_____ᴛᴏ_____

ESTIMATE.

For_____Miles.   Gauge_____

		@ £ s. d.	£
ᴿper yard, with fish plates	Tons	.. ..	..
ᴿspikes, and fang bolts	,,	.. ..	..
S × in. × ft. in. (per mile) ..	No.	.. ..	..
ᴴft. tongues ft. stocks	Sets	.. ..	..
ᴄ with check rails	,,	.. ..	..
ᴿls and trollies	,,	.. ..	..
ᵀurning tank engines (centre balance) ft.	No.	.. ..	..
ᴿing-machines	,,	.. ..	..
ᵛ tons), marked in English	,,	.. ..	..
ᴴrith two brass barrels, fly-wheel, pipes, &c.	,,	.. ..	..
ᵀon plates, with planed ⎫ to hold galls.	,,		
rted on columns, with ⎬ ,, ,,	,,	.. ..	..
leather hose, &c. .. ⎭ ,, ,,	,,		
ᵛngs, engine,. boiler, shafting, lathes and es, fan and smiths' fires, taps and dies, drills, &c.	,,	.. ..	..
S oil and tallow boxes, rope, chain, spun-s, screw jacks, &c., &c.	,,	.. .. .	..
ᴱr fire-box, brass tubes, steel tyres, with bls complete	,,	.. ..	..
ᴿes	Sets	.. ..	..
ᴄns, fitted for luggage and parcels	No.	.. ..	..
ᴄgons, ft. in. × ft. in. × ft. in.; y, cwt.; full, tons	,,	.. ..	..
ᴄgons, ft. in. × ft. in. × ft. in.; y, cwt.; full, tons	,,	.. ..	..
ᴴwheels and axles, axle boxes, springs, ьer set	,,	.. ..	..
ᴺ|lers for ft. span	per pair	.. ..	..
,, ...		.. ..	..
sts ,, ...		.. ..	..

ᵀht.

ᵀht.

ᴬL Cᴏsᴛ of all materials and fittings for   miles (including sleepers)   £_____

,, ,, per mile .. .. £

Tons for Freight.						
Measure-ment.	Dead Weight.	FROM_____TO_____				

ESTIMATE.

Measure-ment.	Dead Weight.	For_____Miles. Gauge_____		£	@ s. d.	£
		s. per yard, with fish plates	Tons	..	..	..
		og spikes, and fang bolts	,,	..	..	..
		in. × in. × ft. in. (per mile) ..	No.	..	..	..
		ft. tongues ft. stocks	Sets	..	..	..
		with check rails	,,	..	..	..
		:ools and trollies	,,	..	..	..
		or turning tank engines (centre balance) ft.	No.	..	..	..
		ghing-machines	,,	..	..	..
		:s (tons), marked in English	,,	..	..	..
		i, with two brass barrels, fly-wheel, pipes, &c.	,,	..	..	..
		t-iron plates, with planed ⎫ to hold galls.	,,			
		ported on columns, with ⎬ ,, ,,	,,	..	..	..
		res, leather hose, &c. .. ⎭ ,, ,,	,,			
		ittings, engine, boiler, shafting, lathes and vices, fan and smiths' fires, taps and dies, ls, drills, &c.	,,	..	..	..
		es, oil and tallow boxes, rope, chain, spun-.ocks, screw jacks, &c., &c.	,,	..	..	..
		pper fire-box, brass tubes, steel tyres, with . tools complete	,,	..	..	..
		icates	Sets	..	..	..
		: vans, fitted for luggage and parcels	No.	..	..	..
		wagons, ft. in. × ft. in. × ft. in. ; npty, cwt. ; full, tons	,,	..	..	..
		wagons, ft. in. × ft. in. × ft. in. ; mpty, cwt. ; full, tons	,,	..	..	..
		of wheels and axles, axle boxes, springs, c., per set	,,	..	..	..
		girders for ft. span .. .. per pair		..	..	..
		,, ,, ...		..	..	..
		d joists ,, ...		..	..	..
		Veight.				
		ement.				
		TOTAL COST of all materials and fittings for miles (including sleepers)				£
		' ,, ,, per mile .. .. £				

Milton Keynes UK
Ingram Content Group UK Ltd.
UKHW022319230424
441667UK00005B/86

9 781019 635445